USE OF LABORATORY ANIMALS IN BIOMEDICAL AND BEHAVIORAL RESEARCH

Committee on the
Use of Laboratory Animals in
Biomedical and Behavioral Research

Commission on Life Sciences
National Research Council

Institute of Medicine

NATIONAL ACADEMY PRESS
Washington, D.C. 1988

NOTICE: The project that is the subject of this report was approved by the Governing Board of the National Research Council, whose members are drawn from the councils of the National Academy of Sciences, the National Academy of Engineering, and the Institute of Medicine. The members of the committee responsible for the report were chosen for their special competences and with regard for appropriate balance.

This report has been reviewed by a group other than the authors according to procedures approved by a Report Review Committee consisting of members of the National Academy of Sciences, the National Academy of Engineering, and the Institute of Medicine.

The National Academy of Sciences is a private, nonprofit, self-perpetuating society of distinguished scholars engaged in scientific and engineering research, dedicated to the furtherance of science and technology and to their use for the general welfare. Upon the authority of the charter granted to it by the Congress in 1863, the Academy has a mandate that requires it to advise the federal government on scientific and technical matters. Dr. Frank Press is president of the National Academy of Sciences.

The National Academy of Engineering was established in 1964, under the charter of the National Academy of Sciences, as a parallel organization of outstanding engineers. It is autonomous in its administration and in the selection of its members, sharing with the National Academy of Sciences the responsibility for advising the federal government. The National Academy of Engineering also sponsors engineering programs aimed at meeting national needs, encourages education and research, and recognizes the superior achievements of engineers. Dr. Robert M. White is president of the National Academy of Engineering.

The Institute of Medicine was established in 1970 by the National Academy of Sciences to secure the services of eminent members of appropriate professions in the examination of policy matters pertaining to the health of the public. The Institute acts under the responsibility given to the National Academy of Sciences by its congressional charter to be an adviser to the federal government and upon its own initiative, to identify issues of medical care, research, and education. Dr. Samuel O. Thier is president of the Institute of Medicine.

The National Research Council was established by the National Academy of Sciences in 1916 to associate the broad community of science and technology with the Academy's purposes of furthering knowledge and of advising the federal government. Functioning in accordance with general policies determined by the Academy, the Council has become the principal operating agency of both the National Academy of Sciences and the National Academy of Engineering in providing services to the government, the public, and the scientific and engineering communities. The Council is administered jointly by both Academies and the Institute of Medicine. Dr. Frank Press and Dr. Robert M. White are chairman and vice-chairman, respectively, of the National Research Council.

This project was undertaken with both public and private sector support. The following agencies of the federal government provided major funding for the study: the Departments of the Air Force, Army, and Navy; the National Institutes of Health; and the National Science Foundation. The following private organizations also provided support: Abbott Laboratories, American Cyanamid Co., American Hoechst Corp., Berlex Laboratories Inc., Bristol-Myers Co., Burroughs Wellcome Co., Ciba-Geigy Corp., E. I. Dupont de Nemours & Co., Marion Laboratories Inc., Pfizer Inc., A. H. Robins Co., Rorer Group Inc., Sandoz Pharmaceuticals Corp., Schering-Plough Corp., Searle Research and Development, Shell Development Co., Sterling Drug Inc., Syntex Corp., and the Upjohn Co.

This work is related to Department of the Navy Grant No. N00014-85-G-0247 issued by the Office of Naval Research. The United States Government has a royalty-free license throughout the world in all copyrightable material contained herein.

Library of Congress Catalog Card Number 88-62248
ISBN 0-309-03839-1

Printed in the United States of America
First Printing, September 1988
Second Printing, January 1989
Third Printing, April 1989
Fourth Printing, September 1989

COMMITTEE ON THE USE OF LABORATORY ANIMALS IN BIOMEDICAL AND BEHAVIORAL RESEARCH

NORMAN HACKERMAN (*Chairman*), Rice University
KURT BENIRSCHKE, University of California, San Diego Medical Center
MICHAEL E. DEBAKEY, Baylor College of Medicine
W. JEAN DODDS, New York State Department of Health, Albany, New York
EDWARD L. GINZTON, Varian Associates, Inc., Palo Alto, California
CARL W. GOTTSCHALK, The University of North Carolina, Chapel Hill
ARTHUR C. GUYTON, University of Mississippi School of Medicine
WILLIAM HUBBARD, The Upjohn Company, Hickory Corners, Michigan
JOHN KAPLAN, Stanford University School of Law
HAROLD J. MOROWITZ, Yale University
CARL PFAFFMANN, The Rockefeller University
DOMINICK P. PURPURA, Albert Einstein College of Medicine
CHRISTINE STEVENS,* Animal Welfare Institute, Washington, D.C.
LEWIS THOMAS, Memorial Sloan-Kettering Cancer Center, New York City
JAMES MCKENDREE WALL, *The Christian Century*, Chicago, Illinois

Staff

JOHN E. BURRIS, *Study Director*
JUNE S. EWING, *Staff Officer*
STEVE OLSON, *Editor*
MARGARET FULTON, *Senior Secretary*
FRANCES WALTON, *Administrative Secretary*

*Did not sign the report.

COMMISSION ON LIFE SCIENCES

Contents

Preface

One of the most difficult issues to confront biological scientists, as well as the society within which they work, is that of the use of laboratory animals in biomedical and behavioral research. What is the ethical relationship of investigators to the animals they use? How may we balance society's desire for the beneficial outcomes of research with the need to protect animals that generally must be used to yield those outcomes? Are there truly effective alternative methods to the use of animals or are these available methods mainly complementary? Are regulations concerning appropriate care of animals too lenient or too strict? The questions are neither easy to answer nor are they new. In 1896, the National Academy of Sciences issued a statement in which it affirmed the need to use animals in medical research in a letter to United States Senator Jacob H. Gallinger (Appendix A). The question arose because of concerns about the treatment of animals in research, and the issues did not differ greatly from those being raised today.

Yet the debate and activity have intensified recently. Research laboratories have been raided to "liberate" animals. The National Institutes of Health has issued sanctions against several institutions for lack of full adherence to animal care regulations. Researchers, who have often failed to present persuasively the case for animal use in research to the public and to politicians, now are becoming increasingly anxious about the limitations placed on their research.

Books have appeared presenting the argument that the "rights" of animals must be considered equal to those of humans.

In September 1985, the National Research Council, through its Commission on Life Sciences and with the collaboration of the Institute of Medicine, appointed a committee to examine concerns about animal use and treatment, benefits derived by humans and animals from research with animals, and current regulatory and self-regulatory guidelines for animal care and use. Care was taken to ensure that the committee membership included scientists from a wide range of disciplines, as well as nonscientists involved in animal welfare, law, and ethics. The diverse backgrounds and interests of the committee provided an opportunity for a wide range of views on the issues to be presented. However, it also meant that on several of these issues it was not possible for the committee to reach unanimity, for example, on the uses of pound animals and coverage of rats, mice, birds, and farm animals used in biomedical research under the Animal Welfare Act. The committee was asked to focus on the use of laboratory animals in research; other uses in testing and education were to be addressed less intensively. This report is the result of the committee's efforts.

The committee met 10 times to collect information, interview knowledgeable persons, and discuss the issues. One meeting was a public forum in which the committee heard a spectrum of views from animal welfare groups, scientists, and anyone else who wished to share their information or opinions. More than 200 persons attended and over 50 individuals representing 40 organizations made oral presentations. In addition, written statements were received from individuals and organizations who did not attend the public meeting. In the course of its work, the committee also invited to other of its meetings about 20 persons who could add to its information on government regulatory and research goals, the philosophies underlying opinions on the appropriateness of animal use, the status of the development of alternative methods to the use of animals, and the experience of other countries in the use of animals and the regulation thereof. The committee benefited from the wisdom of all these contributors and acknowledges its gratitude for their willingness to assist in this study.

The committee had hoped to draw on the results of a new survey of laboratory animal use by the Institute of Laboratory Animal Resources of the National Research Council. Contractual difficulties postponed this survey; thus, discussions of the numbers of animals used are based on other data.

The committee organized itself to address its charge by forming three subcommittees—on societal issues, on regulatory issues, and on scientific issues. The draft reports of these three sections were discussed by the entire committee as it formed its conclusions and recommendations and developed this final report.

No committee dealing with an issue so emotionally charged as this, as diverse in background, and displaying such a vast difference of opinion about its topic could be expected to come easily to a consensus. The issues profoundly affect the performance of scientific research, but they are not themselves scientific. Individual opinion, life experience, and worldview play a large part in determining how any individual approaches the topic. The committee members strove to put aside their personal interests and to address the overarching principal issues in a dispassionate manner. The consensus, as expressed in the conclusions and recommendations of this report, is evidence that the committee was reasonably successful. After discussion on some issues, conclusions were reached that, following further debate, were changed. All had their say, but nevertheless some members wanted to make individual statements. Included at the end of this report are two such statements. These may be useful to help the reader appreciate the depth of feeling and range of individual opinions that exist on this matter among interested people. They also indicate that on some of the issues surrounding the use of laboratory animals, the differences of opinion are too great for a consensus to be reached at this time.

Some feel that the timing of this assignment was inopportune. The scientific community has had little time to work with and adjust to the new regulations that govern animal research, which makes it difficult to assess the impact of the regulatory framework. However, most feel that there is no ideal moment to assess the use of laboratory animals, for, as history has demonstrated, this issue has been under active consideration for well over a century. For example, in the intervening years since the National Academy of Sciences issued its statement in 1896 on the use of laboratory animals, the level of public interest has varied, but the issues and concerns have never disappeared.

We do not expect that this report will end debate about the use of animals in biomedical and behavioral research. That discussion in its modern form has been ongoing for more than a century and is almost continuously at a critical point. We believe, however, that the report provides a carefully reasoned statement on the issues. It also provides a point of departure for further discussion on how to use animals appropriately, while recognizing and being sensitive to the concerns of all segments of our society.

The committee thanks those who have contributed to its work. We are grateful to all who shared their views with us at our public meeting and to those who accepted our invitation to provide information at other committee meetings. A list of this latter group is included as Appendix C. We appreciate the information on the legislative and regulatory framework affecting animal research that was provided by Marcia D. Brody, who served as a consultant.

We wish especially to recognize the efforts of the staff of the Commission on Life Sciences who were instrumental in organizing this effort and in working with the committee throughout the effort. John Burris was a positive influence and a strong and steady guide through the National Research Council report process. June Ewing provided valuable assistance in many areas, and Alvin Lazen brought wisdom and insight. Steve Olson edited the report. The information provided by Wayne Grogan and Dorothy Greenhouse of the NRC's Institute of Laboratory Animal Resources was of great value to us. Barbara Filner of the Institute of Medicine staff also provided useful insights. Mary Frances Walton's cheerful assistance made our work go more easily and our attendance at meetings more enjoyable. Margaret Fulton prepared this manuscript, patiently making the changes required with each revision. We thank them all.

Norman Hackerman, *Chairman*
Committee on the Use of Laboratory Animals in Biomedical and Behavioral Research

Executive Summary

The use of animals in scientific research has been a controversial issue for well over a hundred years. The basic problem can be stated quite simply: Research with animals has saved human lives, lessened human suffering, and advanced scientific understanding, yet that same research can cause pain and distress for the animals involved and usually results in their death. It is hardly surprising that animal experimentation raises complex questions and generates strong emotions.

Animal experimentation is an essential component of biomedical and behavioral research, a critical part of efforts to prevent, cure, and treat a vast range of ailments. As in the past, investigators are using animals to learn about the most widespread diseases of the age, including heart disease and cancer, as well as to gain basic knowledge in genetics, physiology, and other life sciences. Animals are also needed to combat new diseases, of which acquired immune deficiency syndrome (AIDS) is currently the most prominent example. At the same time, behavioral researchers are drawing on animal studies to learn more about such major causes of human suffering as mental illness, drug addiction, and senility.

The recognition that animals are essential in scientific research is critical in making decisions about their use, but these decisions are also made in the broad context of social and ethical values. In this report, the committee addresses these issues and examines how and why animals are used in research and how society oversees that research.

PATTERNS OF ANIMAL USE

Data about the numbers and species of animals used for scientific experimentation in the United States come primarily from two sources: the National Research Council's Institute for Laboratory Animal Resources (ILAR) and the U.S. Department of Agriculture's Animal and Plant Health Inspection Service (APHIS). Though the information from both of these sources is incomplete, it provides a picture of the magnitude of animal experimentation in the United States. In 1983, an estimated 17 to 22 million animals were used for research, testing, and education in the United States. In this case, "animals" includes all vertebrates—namely, mammals, birds, reptiles, amphibians, and fish. The majority of animals used—between 12 million and 15 million—were rats and mice. These quantities are a small fraction of the total of over 5 billion animals used annually for food, clothing, and other purposes in the United States.

A significant portion of the laboratory animals used each year are involved not in research but in testing. Research and testing are not always separable, but testing generally entails the use of animals, primarily rats and mice, to assess the safety or effectiveness of consumer products such as drugs, chemicals, and cosmetics.

The data concerning the numbers of animals used in testing are not complete. Various sources estimate that anywhere from several million to more than half of the approximately 20 million animals used for research and testing in the United States are used for testing. In contrast, the use of animals in education is relatively small (i.e., only an estimated 53,000 animals are used per year in teaching in medical and veterinary schools) and has been declining in recent years.

In general, the data concerning animal use in the United States must be viewed as uncertain. The Office of Technology Assessment has concluded that it is not even possible to tell from the existing data whether the total number of animals used each year is increasing or decreasing. A survey now being planned by ILAR, the fourth in a series of ILAR surveys conducted since 1962, will provide some of this information.

Animal research encompasses a wide range of biomedical and behavioral experiments. One field of behavioral research entails observing animals in colonies that simulate their natural environments. Other animals undergo medical procedures such as surgery to gauge the effectiveness of new techniques. Some are exposed to toxic substances until death or disability results. Others are killed immediately to obtain an essential organ or tissue for further studies. Although long-term survival is sometimes the goal of animal experimentation, most research animals are humanely killed at some point during the course of the research.

BENEFITS DERIVED FROM THE USE OF ANIMALS

The use of animals in biomedical and behavioral research has greatly increased scientific knowledge and has had enormous benefits for human health. For example, in the United States, animal experimentation has contributed to an increase in average life expectancy of about 25 years since 1900. A few examples give an indication of the breadth and variety of these contributions.

- Animals have been used to study cardiovascular function and disease since the early 1600s. Heart-lung machines, which have made open-heart surgery possible, were developed with animals before being used with humans. More than 80 percent of all congenital heart diseases that were formerly fatal can now be cured by surgical treatment based on animal experiments. Similarly, a wide variety of surgical techniques and drug treatments, which have extended life for millions of Americans, were first perfected in animals.
- Studies of the biology of transplantation in animals have made it possible to transfer organs between people. Some 30,000 Americans now alive have transplanted kidneys, which free them from the laborious and uncomfortable dialysis treatments once needed to keep them alive. Other Americans are now alive because of transplanted hearts or livers, or have had their lives immeasurably improved because of skin or cornea transplants. Basic research on transplantation has also contributed greatly to the understanding of immunology, with wide ramifications for the treatment of many diseases.
- Animal research shed light on the nature of polio and has helped to nearly eliminate the disease from the United States. In the early 1900s, researchers succeeded in transmitting the polio virus to monkeys for the first time. In following years, investigators tested various altered or inactivated forms of the virus in monkeys until strains were found that could immunize the monkeys without giving them the disease. This work led to human vaccines that have reduced

the number of cases of paralytic polio in the United States from 58,000 in 1952, at the height of one epidemic, to 4 in 1984.

- Many clinically useful methodologies were first tested on animals before being used with humans. Examples include computed axial tomographic (CAT) scans and magnetic resonance imaging (MRI).
- Animal studies have been essential in probing the functions of the brain in health and disease. Investigators have used animals to understand movement (and the movement dysfunctions caused by such diseases as epilepsy and multiple sclerosis), vision, memory (including the severe memory loss that occurs in 5 percent of persons over the age of 65), drug addiction, nerve cell regeneration, learning, and pain.

The use of animals is important if biomedical research is to continue to lead to the understanding and amelioration of diseases such as cancer, diabetes, and uncontrolled infectious diseases. It will also be essential in efforts to understand and control newly emergent human diseases. For example, researchers have identified viruses in monkeys and other animals that cause diseases in those species similar to AIDS. These animals can therefore act as model systems for the human disease, allowing investigation of possible treatments and vaccines.

Animal research does not only benefit humans. Much animal research also benefits animals, either directly because animal health is the subject of research or indirectly because the same procedures and treatments used in humans can be used in animals. Most of the animals that benefit from this research are domesticated and therefore assist humans in some way—as sources of food and fiber, for instance, or as pets and companions. Vaccines, antibiotics, anesthetics, and other products have improved the lives of countless animals.

ALTERNATIVE METHODS IN BIOMEDICAL AND BEHAVIORAL RESEARCH

Scientists have been and are searching for alternative methods to the use of animals in biomedical and behavioral research for a variety of reasons, including an interest in the welfare of animals, a concern for the increasing costs of purchasing and caring for animals, and because in some areas alternative methods may be more efficient and effective research tools. In current usage, the term "alternative methods" includes replacements for mammals, reductions in the use of animals, and refinements in experimental protocols that lessen the pain of the animals involved.

One way to reduce the use of mammals is to modify experimental protocols so that fewer of them are needed. In the field of testing, for instance, methods have been found to assess toxicity using fewer mammals than were once thought necessary. In addition, in some experimental situations, features of mammals can be modeled by nonmammalian vertebrates (birds, reptiles, amphibians, and fish), invertebrates, plants, organs, tissues, cells, microorganisms, and nonbiological systems. For example, research conducted on the fruit fly *Drosophila* has led to understandings in genetics that apply to all living things, and mathematical models can increase the effectiveness of experiments by defining variables and checking theories, thus making experiments on biological systems more effective and economical. Finally, experimental protocols can be refined to reduce the pain and suffering experienced by laboratory animals. These approaches are all referred to as alternatives.

The search for alternatives to the use of animals in research and testing remains a valid goal of researchers, but the chance that alternatives will completely replace animals in the foreseeable future is nil. Nevertheless, successes have occurred in reducing the numbers of animals used, in developing nonmammalian models, and in refining experimental protocols to reduce the pain experienced by animals, and work continues in this area.

Recognizing the above, the committee recommends that:

- Research investigators should consider possible alternative methods before using animals in experimental procedures.

To enable researchers better to consider alternatives, it is important that they have access to relevant information. The committee therefore recommends that:

- Databases and knowledge bases should be further developed and made available for those seeking appropriate experimental models for use in the design of research protocols.

Furthermore, although the committee's work has focused mainly on research, it recommends that:

- Federal regulatory agencies should move rapidly to accept tests—as such tests become validated—that reduce the number of vertebrates used, insofar as this does not compromise the regulatory mission of an agency and protection of the public.

REGULATORY ISSUES

The laws and regulations governing animal research reflect the broad ethical considerations surrounding the use of animals by humans. The most important federal law affecting animal research in the United States is the Animal Welfare Act. Passed in 1966 and amended in 1970, 1976, and 1985, the act sets minimum standards for handling, housing, feeding, and watering laboratory animals and establishes basic levels of sanitation, ventilation, and shelter from temperature and weather extremes. The law covers those warm-blooded animals designated by the secretary of the U.S. Department of Agriculture, the overseer of the Animal Welfare Act. At present, this includes dogs, cats, nonhuman primates, rabbits, hamsters, guinea pigs, and marine mammals, but not rats, mice, birds, and farm animals used in biomedical research—although rats and mice account for about 85 percent of the animals used in research, education, and testing.

The most recent amendments to the Animal Welfare Act, which took the form of the Improved Standards for Laboratory Animals Act of 1985, added several important provisions to the law. The law requires investigators to consider alternative methods that do not involve animals and to consult with a veterinarian before beginning any experiment that could cause pain. It also requires that dogs receive proper exercise, that primates be provided with environments that promote their psychological well-being, and that all animals used receive adequate presurgical and postsurgical care and pain-relieving drugs. These amendments also require that each registered research facility appoint a committee to monitor animal research in that institution. These committees must include a veterinarian and a person unaffiliated with the research facility to represent the community's interests in animal welfare. Committee members must inspect the facility's animal laboratories twice a year and report deficiencies to the institution for correction. If the deficiencies are not corrected promptly, the U.S. Department of Agriculture must be notified for enforcement, and any funding agency must be informed so that it can decide whether to suspend or revoke grants or contracts to the violator.

A second long-standing, important document affecting animal research in the United States is a product not of the federal government but of the scientific community. In 1963, the Animal Care Panel released the *Guide for Laboratory Animal Facilities and Care.* The *Guide* has been revised five times since then by ILAR, most recently in 1985, and has been renamed the *Guide for the Care and Use of Laboratory Animals* to reflect its broadened scope. Its purpose is to assist investigators and institutions in caring for and using laboratory animals professionally and humanely. It is written in general terms so that it can be used by the wide variety of institutions that conduct experiments using animals.

A number of other government agencies and private organizations have drawn on the *Guide* in establishing standards for animal research. The 1985 Health Research Extension Act, which reauthorized funding for the National Institutes of Health (NIH), requires that researchers receiving funding from NIH adhere to the standards of the *Guide*. In 1986, the Public Health Service (PHS)—which includes NIH, the Food and Drug Administration, the Centers for Disease Control, and the Alcohol, Drug Abuse, and Mental Health Administration—released the most recent revision of its policy statement on the humane care and use of laboratory animals. This, too, requires compliance with the *Guide*. An Interagency Research Animal Committee incorporated the *Guide* by reference in its 1985 "U.S. Government Principles for the Utilization and Care of Vertebrate Animals Used in Testing, Research, and Training." On the nongovernmental side, the American Association for Accreditation of Laboratory Animal Care uses the *Guide* in evaluating the animal facilities of institutions seeking accreditation.

In addition to requiring compliance with the *Guide*, the PHS policy statement and 1985 Health Research Extension Act include several other important statutory and regulatory changes. They require that each institution receiving funds from PHS maintain an Institutional Animal Care and Use Committee (IACUC) to monitor animal research. As with the committees required by the Animal Welfare Act, each IACUC must include one veterinarian and one individual not affiliated with the institution. Investigators who plan to use animals must submit their research protocols to these committees, including a justification for the use of a particular kind of animal and a demonstration that they have considered methods that do not use animals.

The use of animals for research, testing, and education is also regulated in other ways in the United States. For example, the Food and Drug Administration and the Environmental Protection Agency have established Good Laboratory Practices (GLP) regulations that affect the use and care of animals.

Even with this abundance of regulatory activity, self-regulation is the most important determinant of humane treatment of animals. Professional societies have set up guidelines to be followed by their members. In addition, many individual institutions—governmental, academic, and private—have established policies governing animal experimentation and testing. Many institutions now provide information and instruction to animal users on the proper care and handling of research animals. Most important are individual investigators; under the review of their institutional animal committees, they ultimately have the greatest control over and responsibility for how an animal will be cared for and used. At the same time, most scientists acknowledge the need for regulations to set minimum standards and provide for public accountability.

Although humane care and use of laboratory animals characterize the scientific community, there have been from time to time some members of this community who have been found to care inadequately for their animals. The committee believes that the mistreatment or mishandling of animals is not acceptable. Maltreatment and improper care of animals used in research cannot be tolerated, and individuals responsible for such behavior must be subject to censure. Without such punishment, the continued use of animals by all scientists is threatened, as more regulations and restrictions are imposed by legislative and regulatory authorities in response to their perception that scientists who commit abuses are not punished.

Many scientists believe, however, that present regulatory procedures can in some instances be disruptive, in that they may decrease efficiency, increase costs, and slow progress. For instance, obtaining preliminary approval of all research protocols does delay some experiments. On the other hand, protocol review can help the researcher when it provides an opportunity for the scientist's peers to offer advice and assistance. This advice may result in a better-planned experiment that not only improves animal care and minimizes animal pain but also leads to more instructive results. In any case, more extensive regulations may have contributed to the increased expense of animal research, which constrains the research that can be done.

The requirement that investigators strictly comply with the *Guide for the Care and Use of Laboratory Animals* has also raised difficulties. The 1985 Health Research Extension Act essentially imparts the force of law to the *Guide*, but the *Guide* was not written to be a legal document. It was designed to provide for flexibility in interpretation, guided by professional judgment. As such, it has served the community of individuals using laboratory animals well in the more than 20 years since it was first published. Because it is now being used to set minimum standards for inspection, it may in some respects be too rigidly interpreted, as in the requirement for multiple separate areas and rooms for performing aseptic surgery. If the *Guide* is to act as law, it should be carefully examined and redrafted as needed to ensure that its language satisfies the intent, as distinct from the letter, of the law.

In the general area of regulation, the committee recommends the following:

- No additional laws or regulatory measures (except the regulations required by the Improved Standards for Laboratory Animals Act of 1985) affecting the use of animals in research should be promulgated until, based on experience, a careful accounting of the effects of the application of the present body of laws, regulations, and guidelines has been made and evidence of the need for more regulation is available.
- A mechanism should be established for ongoing review of the regulatory framework of federal agencies for animal experimentation. It is essential that research scientists who must abide by this regulatory framework be prominently involved in its assessment. Specifically, the *Guide for the Care and Use of Laboratory Animals* should be reviewed as soon as possible to determine whether revisions are necessary due to new information.
- Federal standards developed by different agencies for the care and use of laboratory animals should be congruent with each other.
- Sufficient federal funds should be appropriated for the inspections required for the enforcement of the Animal Welfare Act.
- Sufficient federal funds should be appropriated for maintenance and improvement of animal facilities to allow individuals and institutions to conduct animal research in compliance with government policies, regulations, and laws. It is important that such funds should be added to ongoing research support.

USE OF POUND ANIMALS

One of the most controversial areas in the current debate involves the use of impounded dogs and cats. The emotions engendered have resulted in the passage of laws by a number of political jurisdictions that prohibit or restrict the release of impounded animals for use in research. These laws create a dilemma: the impounded animals are not released for use in research but are killed by the pound or shelter if not claimed. Each year more than 10 million such animals are destroyed at pounds or shelters, whereas fewer than 200,000 dogs and cats are released from pounds and shelters to scientific establishments for use in research—less than 2 percent of the number that are destroyed.

A prohibition against the use of pound animals also means that more animals are used each year. Instead of using one of the 10 million pound animals that will be destroyed, different animals are bred for use in research.

Whether a pound animal or a "purpose-bred" animal is the appropriate research model depends on the needs of the experiment. Pound animals are seen as having varied genetic backgrounds. In some experiments the genetic variability, because it is much like that found naturally in humans, is an advantage; in other cases it is necessary to know the genetic background of the animal, requiring an animal bred for research. For other experiments it may be necessary to use purpose-bred animals because the health history, physiological status, and age of pound animals are not well enough known to ensure that conditions present in the animals will not interfere with conduct of the experiment.

Twelve states have passed laws that prohibit the release of impounded animals for use in research. In 11 of these states, researchers can use animals impounded in other states, which are legally transported across state lines by dealers. In Massachusetts, a new law that went into effect in 1986 prohibits researchers from using any animals from pounds, no matter where those animals were impounded.

A prohibition against the use of pound animals inevitably increases the costs of animal research because the cost of an animal from a dealer is greater than the cost of a pound animal. If the impounded dogs used each year in research were not available, a substantial additional cost would be incurred from buying replacement dogs from dealers.

In addressing the use of pound animals:

- The committee unanimously recommends that pound animals be made available for research in which the experimental animals are used in acute experiments (i.e., in which the animals remain anesthetized until they are killed). While a majority of the committee supports the appropriate use of pound animals in all experiments, a minority opposes the use of pound animals for chronic, survival experiments.

American society is a pluralistic society in which public policy takes into account many different perspectives. No single ideology or theology governs people's ways of thinking. Similarly, decisions in the United States do not arise unilaterally from authorities. They reflect a consensus within society, as expressed through people's elected representatives.

Some people will continue to contend that animal research should be eliminated. The committee rejects such a view. Indeed, the committee concludes that:

- Humans are morally obliged to each other to improve the human condition. In cases in which research with animals is the best available method to reach that goal, animals should be used.

The committee also recognizes that:

- Scientists are ethically obliged to ensure the well-being of animals used in research and to minimize their pain and suffering.

The committee affirms the principle of humane care of all animals used in research and recommends that:

- All those responsible for the care and use of animals in research should adhere to the principle that these animals be treated humanely.

1
Introduction

Animal experimentation has been a part of biomedical and behavioral research for several millennia; experiments with animals were conducted in Greece over 2,000 years ago. Many advances in medicine and in the understanding of how organisms function have been the direct result of animal experimentation.

Concern over the welfare of laboratory animals is also not new, as reflected in the activities of various animal welfare and antivivisectionist groups dating back to the nineteenth century. This concern has led to laws and regulations governing the use of animals in research and to various guides and statements of principle designed to ensure humane treatment and use of laboratory animals.

HISTORICAL BACKGROUND

Use of Animals in Research

Some of the earliest recorded studies involving animals were performed by Aristotle (384–322 B.C.), who revealed anatomical differences among animals by dissecting them (Rowan, 1984). The Greek physician Galen (A.D. 129–199) maintained that experimentation led to scientific progress and is said to have been the first to conduct demonstrations with live animals—specifically pigs—a practice later extended to other species and termed "vivisection" (Loew,

1982). However, it was not until the sixteenth century that many experiments on animals began to be recorded. In 1628, William Harvey published his work on the heart and the movement of blood in animals (French, 1975). In the 1800s, when France became one of the leading centers of experimental biology and medicine—marked by the work of such scientists as François Magendie in experimental physiology, Claude Bernard in experimental medicine, and Louis Pasteur in microbiology and immunology—investigators regularly used animals in biomedical research (McGrew, 1985).

Research in biology progressed at an increasing pace starting around 1850, with many of the advances resulting from experiments involving animals. Helmholtz studied the physical and chemical activities associated with the nerve impulse; Virchow developed the science of cellular pathology, which led the way to a more rational understanding of disease processes; Pasteur began the studies that led to immunization for anthrax and inoculation for rabies; and Koch started a long series of studies that would firmly establish the germ theory of disease. Lister performed the first antiseptic surgery in 1878, and Metchnikoff discovered the antibacterial activities of white blood cells in 1884. The first hormone was extracted in 1902. Ehrlich developed a chemical treatment for syphilis in 1909, and laboratory tissue culture began in 1910. By 1912, nutritional deficiencies were sufficiently well understood to allow scientists to coin the word "vitamin." In 1920, Banting and Best isolated insulin, which led to therapy for diabetes mellitus. After 1920, the results of science-based biological research and their medical applications followed so rapidly and in such numbers that they cannot be catalogued here.

Concerns over Animal Use

The first widespread opposition to the use of animals in research was expressed in the nineteenth century. Even before this, however, concern had arisen about the treatment of farm animals. The first piece of legislation to forbid cruelty to animals was adopted by the General Court of Massachusetts in 1641 and stated that "No man shall exercise any tyranny or cruelty towards any brute creatures which are usually kept for man's use" (Stone, 1977). In England, Martin's Act was enacted in 1822 to provide protection for farm animals. In 1824, the Society for the Prevention of Cruelty to Animals (SPCA) was founded to ensure that this act was observed. In 1865, Henry Bergh brought the SPCA idea to America (Turner, 1980).

He was motivated not by the use of animals in research but by the ill-treatment of horses that he observed in czarist Russia.

In the second half of the nineteenth century, concerns for the welfare of farm animals expanded to include animals used in scientific research. The antivivisectionist movement in England, which sought to abolish the use of animals in research, became engaged in large-scale public agitation in 1870, coincident with the development of experimental physiology and the rapid growth of biomedical research. In 1876, a royal commission appointed to investigate vivisection issued a report that led to enactment of the Cruelty to Animals Act. The act did not abolish all animal experimentation, as desired by the antivivisection movement. Rather, it required experimenters to be licensed by the government for experiments that were expected to cause pain in vertebrates.

As animal experimentation increased in the United States in the second half of the nineteenth century, animal sympathizers in this country also became alarmed. The first American antivivisectionist society was founded in Philadelphia in 1883, followed by the formation of similar societies in New York in 1892 and Boston in 1895. Like their predecessors in England, these groups sought to abolish the use of animals in biomedical research, but they were far less prominent or influential than the major animal-protection societies, such as the American SPCA, the Massachusetts SPCA, and the American Humane Association (Turner, 1980).

Unsuccessful in its efforts toward the end of the nineteenth century to abolish the use of laboratory animals (Cohen and Loew, 1984), the antivivisectionist movement declined in the early twentieth century. However, the animal welfare movement remained active, and in the 1950s and 1960s its increasing strength led to federal regulation of animal experimentation. The Animal Welfare Act was passed in 1966 and amended in 1970, 1976, and 1985. Similar laws have been enacted in other countries to regulate the treatment of laboratory animals (Hampson, 1985).

Concern over the welfare of animals used in research has made itself felt in other ways. In 1963, the Animal Care Panel drafted a document that is now known as the *Guide for the Care and Use of Laboratory Animals* (National Research Council, 1985a). As discussed in Chapter 5, the *Guide* is meant to assist institutions in caring for and using laboratory animals in ways judged to be professionally and humanely appropriate. Many professional societies and public and private research institutions have also issued guidelines and statements on the humane use of animals; for example, the American Physiological Society, the Society for Neuroscience, and the American Psychological Association.

PRESENT SITUATION

Despite the long history of concern with animal welfare, the treatment and use of experimental animals remain controversial. In recent years a great expansion of biomedical and behavioral research has occurred. Simultaneously, there has been increased expression of concern over the use of animals in research. Wide publicity of several cases involving the neglect and misuse of experimental animals has sensitized people to the treatment of laboratory animals. Societal attitudes have also changed, as a spirit of general social concern and a strong belief that humans have sometimes been insensitive to the protection of the environment have contributed to an outlook in which the use of animals is a subject of concern.

Of course, any indifference to the suffering of animals properly gives rise to legitimate objections. From time to time some few members of the scientific community have been found to mistreat or inadequately care for research animals. Such actions are not acceptable. Maltreatment and improper care of animals used in research cannot be tolerated by the scientific establishment. Individuals responsible for such behavior must be subject to censure by their peers. Out of this concern that abuse be prevented, organizations have emerged to monitor how laboratory animals are being treated, and government agencies and private organizations have adopted regulations governing animal care and use.

Discussions about laboratory animal use have also been influenced in recent years by the emergence of groups committed to a concept termed "animal rights." Some of these groups oppose all use of animals for human benefit and any experimentation that is not intended primarily for the benefit of the individual animals involved. Their view recognizes more than the traditional interdependent connections between humans and animals: It reflects a belief that animals, like humans, have "inherent rights" (Regan, 1983; Singer, 1975).

Their use of the term "rights" in connection with animals departs from its customary usage or common meaning. In Western history and culture, "rights" refers to legal and moral relationships among the members of a community of humans; it has not been applied to other entities (Cohen, 1986). Our society does, however, acknowledge that living things have inherent value. In practice, that value imposes an ethical obligation on scientists to minimize pain and distress in laboratory animals.

Our society is influenced by two major strands of thought: the Judeo-Christian heritage and the humanistic tradition rooted in Greek philosophy. The dominance of humans is accepted in both traditions. The Judeo-Christian notion of dominance is reflected in the passage in the Bible that states (Genesis 1:26):

> And God said, Let us make man in our image, after our likeness; and let them have dominion over the fish of the sea, and over the fowl of the air, and over the cattle, and over all the earth, and over every creeping thing that creepeth upon the earth.

However, the Judeo-Christian heritage also insists that dominance be attended by responsibility. Power used appropriately must be used with the morality of caring. The uniqueness of humans, most philosophers agree, lies in our ability to make moral choices. We have the option to decide to dominate animals, but we also have a mandate to make choices responsibly to comply with the obligations of stewardship.

From tradition and practice it is clear that society accepts the idea of a hierarchy of species in its attitudes toward and its regulation of the relationships between humans and the other animal species. For example, animals as different as nonhuman primates, dogs, and cats are given special consideration as being "closer" to humans and are treated differently from rodents, reptiles, and rabbits.

Most individuals would agree that not all species of animals are equal and would reject the contention of animal rights advocates who argue that it is "speciesism" to convey special status to humans. Clearly, humans are different, in that humans are the only species able to make moral judgments, engage in reflective thought, and communicate these thoughts. Because of this special status, humans have felt justified to use animals for food and fiber, for personal use, and in experimentation. As indicated earlier, however, these uses of animals by humans carry with them the responsibility for stewardship of the animals.

Several recent surveys have examined public opinion about the use of laboratory animals in scientific experimentation (Doyle Dane Bernbach, 1983; Media General, 1985; Research Strategies Corp., 1985). Most of the people interviewed want to see medical research continued, even at the expense of animals' lives. Beyond that, people's thoughts about animal use depend on the particular species used and/or on the research problem being addressed. Almost all people support the experimental use of rodents. Support for the use of dogs, cats, and monkeys is less, and people clearly would prefer that rodents be used instead. Most people polled believe that animals used in research are treated humanely.

The next two chapters examine the ways in which animals are used in the United States and the benefits that have been derived from the use of experimental animals. After a discussion of alternative methods in the use of laboratory animals (Chapter 4), the report discusses the regulatory issues surrounding animal use (Chapter 5) and the use of animals from pounds and shelters (Chapter 6). Chapter 7 contains the committee's recommendations.

2
Patterns of Animal Use

Animals are used for a variety of purposes in the United States—for food and other products; in sports and entertainment; for companionship; for the production of enzymes, hormones, and other biological products; and in research, testing, and education. The largest use of animals is in food and fiber production, accounting for over 5 billion vertebrates each year (U.S. Department of Agriculture, 1985). An estimated 110 million dogs and cats are household pets in the United States. Between 17 million and 22 million animals are estimated to be used annually in the United States in research, education, and testing. About 85 percent of these are rats and mice, and less than 2 percent are cats, dogs, and nonhuman primates (Office of Technology Assessment, 1986).

Animals are used in research to improve the health and welfare of humans and animals and to gain basic knowledge that cannot be gained in other ways. Research conducted on animals varies widely in its impact on the animal subjects themselves. One field of behavioral research consists of observations of animals living in colonies that simulate their natural environments but with adequate food supplies and no predators. In some research projects, animals are subjected to experimental procedures and then receive supportive care, because their long-term survival and the validation of methods are the goals of treatment (examples include the development of organ transplantation and chronic toxicology). Some research animals are subjected to toxic substances and painful procedures until they are disabled or die, as when determining the lethal dose of radiation used in cancer therapy. Some are killed to obtain an essential organ, such as the liver, to be used in further studies. Others are anesthetized, subjected to an experimental procedure, and killed without regaining consciousness.

Not only is there considerable variation in how animals are used, but there is variation in how many and what types of animals are used in experiments.

NUMBERS OF ANIMALS USED

In 1952 the National Research Council established the Institute for Laboratory Animal Resources (ILAR) to serve as a coordinating agency and an information resource on the use of laboratory animals. In 1962, 1968, and 1978, ILAR conducted major surveys of laboratory animal facilities and resources, with the results of the 1978 survey being published by the U.S. Department of Health and Human Services (National Research Council, 1980). The 1968 and 1978 ILAR surveys included most of the entities that use animals in biomedical research, including nonprofit, commercial, military, and federal organizations. ILAR is currently planning a fourth survey.

The Animal and Plant Health Inspection Service (APHIS) of the U.S. Department of Agriculture (USDA) also collects data on the extent of animal use. Each year APHIS prepares an Animal Welfare Enforcement Report, which summarizes the annual reports filed with APHIS by registered research facilities that use animals in research (U.S. Department of Agriculture, 1972–1987). All registered research facilities are required to submit these reports. Institutions are not required to report on their use of rats, mice, birds, and domestic farm animals used for research, but the annual report form has space for voluntary reporting on the use of rats and mice.

Table 1 summarizes information from the ILAR and APHIS surveys and from estimates prepared by Health Designs, Inc., for the Office of Technology Assessment (1986). As demonstrated by the table, data from various sources show a lack of consistency. It should be noted that a considerable decrease was observed between 1967 and 1978 in the numbers of animals used as measured by ILAR. Recent annual reports from APHIS, however, have shown that the total number of animals used in experimentation (excluding rats, mice, birds, and wild animals) increased from 1,571,693 in 1983

TABLE 1 Various Estimates of the Number of Animals Used in the United States

Group/Species	ILAR, 1967	ILAR, 1978	APHIS,[a] 1982	Health Designs (est.), 1983	APHIS,[b] 1983	APHIS,[b] 1984	APHIS,[b] 1985	APHIS,[b] 1986	APHIS,[b] 1987
No. of reporting institutions	1,371	1,252	1,127	--	1,005	1,108	1,105	1,126	1,183
Mice	22,772,300	13,413,813	6,889,744	8,500,000					
Rats	6,131,000	4,358,766	2,725,814	3,700,000					
Hamsters	785,900	368,934	417,267	454,479	337,023	437,123	414,460	370,655	416,002
Guinea pigs	613,300	426,665	497,860	521,237	485,048	561,184	598,903	462,699	538,998
Other rodents	60,500	79,993							
Rabbits	504,500	439,986	547,312	509,052	466,810	529,101	544,621	521,773	534,385
Cats	99,300	54,908	59,961	55,346	53,344	56,910	59,211	54,125	50,145
Dogs	262,000	183,063	194,867	182,425	174,542	201,936	194,905	176,141	180,169
Other carnivores	9,100	4,990							

Ungulates	106,200	144,595							
Nonhuman primates	57,700	30,323	54,565	59,336	54,926	55,338	57,271	48,540	61,392
Birds	2,070,500	450,352		100,000[c]					
Amphibians				500,000[c]					
Fish				4,000,000[c]					
Subtotal, excluding rats and mice					1,571,693	1,841,592	1,869,371	1,633,933	1,801,091
Total	33,472,300	19,956,388	11,387,390	18,581,875					

NOTE: APHIS = Animal and Plant Health Inspection Service; ILAR = Institute for Laboratory Animal Resources.

[a]Data were compiled by Health Designs, Inc. (Rochester, N.Y.) from research facilities' annual reports. The data for rats and mice come from voluntary reports of the use of these species.
[b]Data were obtained from Animal Welfare Enforcement Reports to Congress for the year given. They do not include any numbers for rats and mice.
[c]The estimates stated are the highest value of a rough range.

SOURCE: Adopted from Office of Technology Assessment, 1986, Table 3.5.

to 1,633,933 in 1986 (U.S. Department of Agriculture, 1972–1987). The Office of Technology Assessment (1986), in evaluating all the data, has concluded that the available data are too imprecise to allow any conclusions to be made regarding recent trends in overall animal use. The ILAR survey being planned will provide more current information on animal use.

USE OF ANIMALS IN RESEARCH BY THE FEDERAL GOVERNMENT

The federal government is a major user of research animals. Specifically, the following departments and agencies use animals for intramural research and testing (Office of Technology Assessment, 1986).

- *The U.S. Department of Agriculture* conducts research with animals to improve animal health and the quality of animal products, such as food and fiber.
- *The U.S. Department of Defense* conducts experimental research in a wide variety of areas, with animals being used by the Air Force, the Army, the Navy, the Uniformed Services University of the Health Sciences, the Defense Nuclear Agency, and the Armed Forces Institute of Pathology.
- *The U.S. Department of Energy* conducts research on the health and environmental effects of energy technologies and programs. Most of this research takes place at the privately managed national laboratories—such as Brookhaven National Laboratory, Oak Ridge National Laboratory, and the Pacific Northwest Laboratories—and through contracts and grants to scientists employed at universities and other research facilities.
- *The U.S. Department of Health and Human Services* carries out intramural animal research or testing within four of its components: the National Institutes of Health (NIH); the Food and Drug Administration (FDA); the National Institute on Drug Abuse (NIDA), which is part of the Alcohol, Drug Abuse, and Mental Health Administration (ADAMHA); and the National Institute for Occupational Safety and Health (NIOSH), which is part of the Centers for Disease Control. NIH is the largest of these four components and uses more animals than any other federal department or agency.
- *The U.S. Department of the Interior*, in cooperation with state and private organizations, conducts research and education programs to improve fish and wildlife resource management.

- *The U.S. Department of Transportation* conducts research on transportation safety using animals under the authority of the Hazardous Transportation Act of 1974 and the National Traffic and Motor Vehicle Safety Act of 1966.
- *The Consumer Product Safety Commission* (CPSC) conducts tests to determine the toxic potential of consumer products.
- *The Environmental Protection Agency* (EPA) performs research involving animals under the statutory and regulatory authority of the Toxic Substances Control Act and the Federal Insecticide, Fungicide, and Rodenticide Act.
- *The National Aeronautics and Space Administration* (NASA) conducts research with animals to acquire knowledge that can be used to protect the health of astronauts, both during their missions in space and after their return to earth.
- *The Veterans Administration* (VA) uses animals in its research and development divisions and in its education programs.

The Office of Technology Assessment (1986) has estimated that the total federal use of animals in 1983 was 1.6 million, with about 90 percent of these animals being rats and mice.

USE OF ANIMALS IN EDUCATION

The number of animals used in education is unknown, but most observers think that it is relatively small. For example, an estimated 53,000 animals are used annually for teaching in medical and veterinary schools (Office of Technology Assessment, 1986). However, animal use in high schools and colleges might be most people's only contact with laboratory animals, making it an important determinant of how the public feels about such use. This topic is outside the charge of the committee, but the recent report by the Office of Technology Assessment (1986) examines the issue in some detail.

USE OF ANIMALS IN TESTING

Animals are used extensively to test the safety and efficacy of compounds produced by the chemical, cosmetic, and drug industries. The use of so many animals, particularly rats and mice, in testing cannot be ignored even though the committee was charged primarily with looking at the use of animals in research. Government regulatory agencies, such as FDA, EPA, CPSC, and the Occupational Safety and Health Administration (OSHA), often explicitly require the use of animals in testing. A list of some commonly used tests follows (Office of Technology Assessment, 1986). Descriptions of possible alternative methods can be found in Chapter 4.

- *Acute toxicity tests* consist of single doses at concentrations high enough to produce toxic effects or death. They are often used to screen substances for relative toxicity. The LD_{50}, which is the dose of a test substance at which half the test animals can be expected to die, is one such test.
- *Eye and skin irritation tests*, which usually consist of a single exposure, are generally used to develop warnings for handling and to predict the toxicity of accidental exposure. The most common method used to test eye irritation is the Draize test, in which a test substance is applied to one eye of an adult rabbit, with the untreated eye serving as a control (Draize et al., 1944).
- *Repeated-dose chronic toxicity tests* entail repeated exposures to substances for periods of two weeks to more than a year to determine the possible effects of long-term exposure. Rats are most commonly used for these tests.
- *Carcinogenicity tests* involve repeated exposures to substances for most of an animal's lifespan to detect possible human carcinogens.
- *Developmental and reproductive toxicity tests* consist of a variety of procedures to determine the potential of foreign substances to cause infertility, miscarriages, and birth defects. Rats and rabbits are the most commonly used animal subjects.
- *Neurotoxicity tests* use a variety of doses and exposures to determine toxic effects on the nervous system. Toxic end points include behavioral changes, lack of coordination, motor disorders, and learning disabilities in animals.
- *Mutagenicity tests* include a variety of methods for determining whether genetic material of germ or somatic cells has been changed.
- *Biological screening tests* investigate the biological activity of organic compounds. Animals may be used in these tests depending on the type of biological activity being investigated.

Most of the above-mentioned tests require the use of large numbers of animals. However, as mentioned earlier, the number of animals used in testing is not known. Most testing is thought to be conducted in private commercial establishments that use primarily rats and mice, which under current regulations are not subject to the reporting requirements of the Animal Welfare Act. A recent estimate of the total number of animals used in testing was "several" million (Office of Technology Assessment, 1986). Another report (Theta Corporation, 1986) estimated that the use of animals in testing and industrial research is considerably greater than that, with organizations outside of government and academia accounting for over 75 percent of the estimated 22 million laboratory animals used annually. Of these animals, rodents by far are used in the greatest numbers.

NEW TECHNOLOGIES AND FUTURE LABORATORY USE OF ANIMALS

The new and rapidly expanding field of biotechnology will have an impact on the species and numbers of laboratory animals used, but it is too early to predict precisely its ultimate effects. In some cases, the number of animals used might be reduced as biotechnology provides new testing methods acceptable to governmental regulatory authorities. In other cases, biotechnology might cause a need for more animals as well as shifts in the relative numbers of various species of animals used. At present, the biotechnology industry in the United States purchases an estimated 11 percent of all laboratory rodents sold, about 5 percent of the swine, and about 2 percent of the rabbits and dogs, but few primates or cats (Theta Corporation, 1986).

Several effects of biotechnology can already be seen. Rabies virus is widely distributed in nature. It was initially studied by infecting live laboratory animals with the virus, which led to vaccines produced using live animals. Recently, new diagnostic tests have been developed that use monoclonal antibodies produced by cell cultures, and vaccines are being produced with recombinant DNA technology (Freiherr, 1986). These changes have greatly reduced the use of animals for this purpose.

Proteins such as growth hormone and insulin can now be made using bioengineering techniques. Although this method of production will not eliminate the use of animals, it may reduce the number used per product, because safety tests can then be performed with larger batches of a uniform product.

The increasing sophistication in determining molecular structure and using it to predict biochemical function may reduce the use of animals. Scientists can use advances in technology to determine the active sites of molecules and even the attachment sites of viruses.

Such information may permit drug synthesis to proceed in a more directed fashion. New compounds developed in this way will still require safety and efficacy testing in animals. Animals will also still be needed for the validation of predicted results.

The numbers of particular animals used could change. For example, more mice might be used, because transgenic mice produced by the microinjection of DNA into fertilized mouse eggs constitute a powerful system for the study of specific genes (Bieberich and Scangos, 1986).

SUMMARY

No comprehensive data on the use of animals for research, testing, and education in the private sector are available, and trends in this use are difficult to gauge. Federal in-house use amounts to about 1.6 million animals, or less than 10 percent of the estimated 17 million to 22 million animals used annually for research, education, and testing in the United States. A uniform system of reporting, while costly, would help to determine more accurately the numbers of animals used in research, which would make it possible to assess the impact of policy on trends in animal use. Animals are used extensively in testing the safety and efficacy of compounds produced by the chemical, cosmetic, and drug industries. Commonly used tests include those for acute toxicity, eye and skin irritation, repeated-dose chronic toxicity, carcinogenicity, developmental and reproductive toxicity, neurotoxicity, mutagenicity, and biological screening. Future technologies might afford ways of reducing animal use, or they might lead to a need for more animals or to shifts in the relative numbers of different species used.

3
Benefits Derived from the Use of Animals

Animal studies have been an essential component of every field of medical research and have been crucial for the acquisition of basic knowledge in biology. In this chapter a few of the contributions of such studies in biomedical and behavioral research will be chronicled. These descriptions should be viewed within the context of the vast improvements in human health and understanding that have occurred in the past 150 years. For example, since 1900 the average life expectancy in the United States has increased by 25 years (U.S. National Center for Health Statistics, 1988). This remarkable increase cannot be attributed solely to animal research, as much of it is the result of improved hygiene and nutrition, but animal research has clearly been an important contributor to improved human health.

Despite the many advances and the projected results that will come through the use of animals, some individuals question the value of using animal models to study human disease, contending that the knowledge thus gained is insufficiently applicable to humans. Although experiments performed on humans would provide the most relevant information (and are used in clinical research conducted on humans when appropriate), it is not possible by commonly accepted ethical and moral standards or by law to perform most experiments on humans initially. It is true that not every experiment using animals yields immediate and practical results, but the advances that will be described in this chapter provide evidence that this means of research has contributed enormously to the well-being of humankind.

POLIO

As a result of the acquisition of information and the development of techniques achieved through the use of animals, poliomyelitis is no longer a major threat to health in the United States. The number of cases of paralytic polio in the United States has declined as a result of vaccinations from 58,000 in 1952 to only 4 in 1984 (Office of Technology Assessment, 1986). Unfortunately, polio is still a major threat to health where the vaccine is not used. Indeed, in a number of African, Asian, and South American countries, the incidence of the disease has been rising, despite the availability of the vaccine (Cockburn and Drozdov, 1970). An estimated 500,000 cases occur around the world each year (Salk, 1983).

The use of rhesus monkeys for the study of polio began when Landsteiner and Popper (1909) showed that injection of spinal cord material from patients dying of polio caused paralysis in the animals. Flexner and Lewis (1909) promptly confirmed this result. To learn how to immunize monkeys to protect them against infection, researchers first used live virus, then formalin-inactivated virus from infected brain suspensions, and eventually modified live virus. A major breakthrough occurred when Enders, Weller, and Robbins (1949) showed that the virus could be propagated in cultured cells of nonneural origin. That set the stage for mass production of viruses that could be made into formalin-inactivated Salk vaccine or the modified live-virus Sabin vaccine (Salk, 1983).

Although the use of monkeys in polio research has decreased considerably, they are still essential to the production of both live and killed polio vaccines, which are routinely produced in monkey kidney cell cultures. The live vaccine is tested for neurovirulence in monkeys, and the killed vaccine is routinely tested for safety in monkeys.

ACQUIRED IMMUNE DEFICIENCY SYNDROME

The recent emergence of acquired immune deficiency syndrome (AIDS) as a major health threat exemplifies not only the unpredictability of research needs, but also the criticality of animals in research. The similarity of simian AIDS, identified in rhesus monkeys at two primate centers, to human AIDS has allowed the disease in monkeys to serve as a model for the human disease. In monkeys, the virus that causes the disease has been isolated, infectibility studies have been conducted, and some experiments have provided preliminary indications of the possibility of vaccine development. This animal model might prove useful for testing the efficacy and safety of vaccines and therapeutic agents developed to prevent or treat the human disease (Institute of Medicine, 1986).

Recently, a new virus called feline T-lymphotropic lentivirus has been discovered. It resembles morphologically the human immunodeficiency virus (HIV) that causes AIDS, although differing antigenically, and causes a disease naturally in cats similar to AIDS. Thus, infected cats might prove useful as animal models for the study of certain aspects of human AIDS (Pedersen et al., 1987).

TRANSPLANTATION

The transplantation of skin, corneas, and various internal organs could not have become a safe and standard procedure without the knowledge of the biology of transplantation immunology acquired through the use of experimental animals. Some 30,000 Americans now alive have transplanted kidneys, and others survive with transplanted hearts and livers or retain their sight because of corneal transplants.

The treatment of burn victims was of particular importance to the British during World War II, and British biologist P. B. Medawar (1944) undertook to find relief for them through the transplantation of skin. For one of his models, he used freemartin cattle. A freemartin is a sexually maldeveloped female calf that is born as a twin of a normal male calf; male hormones that reach it through placental vessels usually make it sterile (Lillie, 1917). Experimentation showed that skin and other tissues could be transplanted with good, lasting success between the male and freemartin twins at any stage in their lives (Anderson et al., 1961). They were "tolerant" of each other's tissues because of prenatal exposure to each other's tissue antigens. Medawar and his colleagues sought to induce such tolerance in newborn mice. When newborns received skin transplants or received bone marrow from unrelated animals, they became forever "tolerant" of the new tissue (Brent et al., 1976). That discovery signaled a new era in immunology, with wide ramifications for health and the treatment of disease not only in humans, but also in animals.

Through a systematic study of the surface immune markers of specially bred strains of mice, Snell and Benacerraf provided the basis for much of the understanding that has led to the success of organ transplantation (Benacerraf, 1981).

In the past, young women with chronic pyelonephritis, patients with genetic polycystic disease, and people suffering from the aftermath of streptococcal infections were all vulnerable to chronic renal failure and death. Those people benefited from the invention of "artificial kidneys," which periodically washed blood and removed poisonous substances from it. The recipients of the benefit, however, had to undergo frequent, laborious, and uncomfortable procedures and had to rely on hospitals and mechanical devices.

The first extensive work with renal transplantation was reported in 1955 (Hume et al., 1955). At first, transplanted kidneys were rejected unless they were exchanged between identical twins. However, studies in dogs showed that administration of the drug 6-mercaptopurine after transplantation would prolong the survival of a transplanted organ from an unrelated person. This use of immunosuppressants ushered in the modern era of transplantation (Starzl and Holmes, 1964). These compounds, having been studied first in animals and proved to be effective, are now used in human transplant recipients.

The study of tissue antigens proceeded at the same time as transplantation work, first in mice and then in humans. Inbred (isogeneic) strains of mice had been created by repeated brother–sister matings. Ultimately, these strains became genetically identical, and the exchange of tissues and organs became possible. In the study of minor genetic differences between such strains, it became clear that some genes specify the cell-surface structures responsible for tissue recognition and rejection. "Transplantation antigens" can now be identified by tissue typing, and the most appropriate donors can be chosen for transplantation in both humans and animals.

A second revolution in transplantation was ushered in by the development of cyclosporin. This immunosuppressive agent was first used successfully in humans in 1983, after five years of toxicity and efficacy testing in mice, rats, and other animals. Since it became available for heart transplantation, survival after transplantation has improved significantly (Kupiec-Weglinski et al., 1984). Further progress is now occurring with monoclonal antibodies that seem to immobilize the cell-surface markers responsible for recognition and rejection. The hope is that such monoclonal antibodies, which have been developed and maintained in animals, will make it unnecessary to resort to complete immunosuppression of a transplant recipient. This would reduce the occurrence of infection and increase the rates of survival of transplanted organs.

CARDIOVASCULAR-RENAL SYSTEMS

Dogs have traditionally been used in cardiovascular-renal studies because of their relatively large size, which facilitates experimental procedures. For example, an early model of hypertension was produced by partially occluding the renal artery in dogs. Studies of renal function that use clearance techniques in unanesthetized animals are most often done in dogs. In the last two decades, however, some mutant rats have proved exceedingly valuable as animal models of human disease. The Brattleboro rat is an excellent example. It has diabetes insipidus and must drink 70 percent of its body weight in water each day. It cannot produce vasopressin, a hormone that plays an essential role in the kidneys' ability to regulate water excretion and blood pressure. Research on the Brattleboro rat has greatly increased our understanding of vasopressin's role in kidney and cardiovascular function, and that understanding might lead to the development of better drugs (and drugs with fewer side effects) for the treatment of clinical disorders (Sokol and Valtin, 1982).

The development of open-heart surgery is but one of many examples of the value of using laboratory animals. Working with cats and dogs, Gibbon built the forerunner of the present-day heart-lung machine (Deaton, 1974), which makes open-heart surgery possible. His research in the early 1930s included clamping off more and more of an animal's vasculature and detouring its blood through the heart-lung machine. The machine was further improved by the incorporation of a roller pump developed by DeBakey (DeBakey and Henly, 1961), which allowed the entire circulation to be shunted through the machine, which added oxygen to the animal's blood. The pump was first used and perfected in the animal laboratory and is now a standard, essential component of the heart-lung machine. As a result of these developments, more than 80 percent of infants born with congenital cardiac abnormalities now can be treated surgically and can lead normal lives.

Replacement of heart valves and segments of large arteries in the treatment of valvular heart disease was made feasible by dog studies done in the late 1940s and early 1950s (Gay, 1984). Before diseased heart valves could be replaced in patients, scientists had to study their design and use in animals. As with so many other drugs and operations, physicians and surgeons would not consider applying them to patients until they had proved safe and effective in animals, nor would the public accept them until their safety was proved. Each decade since then has seen improvements in the design, installation, and performance of these valves and other prosthetic devices. Because the ideal valve has not yet been developed, research is still in progress in many laboratories to further improve its capacities.

NERVOUS SYSTEM

The human brain is a structure of extraordinary complexity. Each of its 200 billion neurons (nerve cells) makes a few thousand to several hundred thousand connections with other neurons, muscles, or glands. Neurons use large amounts of metabolic energy to carry out a host of functions: the generation and conduction of impulses; the synthesis, transport, secretion, and uptake of transmitters; and the modification of structure and synaptic efficacy in response to activity and environmental perturbations (Kandel and Schwartz, 1985).

Many basic aspects of neuronal development can be studied in cell and tissue cultures, in brain slices, and in simple invertebrate neuronal systems. Computer simulations and noninvasive human studies can also provide important data on fundamental mechanisms of learning and memory. Yet there is no adequate substitute for animal studies in attempts to understand the complex behavioral and cognitive functions of the brain in health and disease.

Movement and Function

Our understanding of the nervous system and approaches to rational therapy of its disorders could not have come about without animal studies initiated by the physiologist Charles Sherrington (Eccles and Gibson, 1979). His studies on reflex mechanisms of the spinal cord in cats were continued by Eccles (1957), who described how excitatory and inhibitory processes work in the central nervous system. Today, neurosurgeons can remove some brain tumors with minimal damage to the motor system in part because scientists such as Sherrington discovered that localized electrical stimulation of the exposed brain of the dog could elicit discrete movements of the limbs.

Neurologists and neurosurgeons now examine electrical signals from the brain to diagnose and treat epilepsy, study levels of consciousness, localize brain tumors, diagnose multiple sclerosis, and study learning disabilities in children. Moreover, the applications of such essential tools for diagnosis and therapy as computed axial tomographic (CAT) scans and magnetic resonance imaging (MRI) were developed with research animals (Kandel and Schwartz, 1985).

Behavior

The study of the nervous system and behavior is one of the major frontiers of modern science. A good deal is known about the anatomy and physiology of the brain and nervous system, but much remains to be learned about it as an organized assemblage of neurons and about how it is affected by environmental stimulation. The following examples provide an idea of how animals are used in studies of such subjects.

Postnatal Development of the Visual Cortex and the Influence of Environment

Hubel and Wiesel shared the Nobel Prize in 1981 for their studies of vision in cats and monkeys, including the development of visual functions in young animals (Barlow, 1982). The visual cortex of monkeys is not fully developed at birth; nerve cells are still growing and making connections with other nerve cells. In this process, normal development depends on visual stimulation during a critical period in early postnatal life.

As in humans, each eye of a monkey sees a slightly different view of the same object; normal binocular vision gives the impression of depth. If early in postnatal life one eye is occluded, the nerve cells for that eye in the visual cortex do not develop normally. Most of the nerve cells become responsive only to the open eye, as shown in recordings from cells of the visual cortex of anesthetized animals. In normal development, the visual cortex consists of alternating bands of reactive neurons from the right and left eyes; but in a monkey with an occluded eye, the regular alternation is weakened, and most neurons react only to the normal eye. These anatomical and physiological changes are the basis of blindness in the occluded eye.

Children with congenital cataracts or clouding of the ocular media for other reasons demonstrate a similar dependence of human vision on visual stimulation. Testing after restoration of normal vision has shown that the acuity of the previously occluded eye is reduced; the earlier in life the eye was occluded, the greater the degree of deficit. Animal experiments have also shown that correction of strabismus (squint) by surgery should be performed early in, or certainly before the end of, the critical period of eye-brain development to ensure normal vision (Wiesel, 1982).

The close correlation between the effects of visual deprivation observed in animals and the effects observed in the clinic suggests that they are based on similar physiological mechanisms. This correlation has been helpful in developing appropriate measures of prevention and treatment of neural eye disorders.

Memory

Another subject of behavioral research is memory. An estimated 5 percent of people over the age of 65 have severe limitations or even failures of memory and cognition; another 10 percent of the people over 65 have mild to moderate cognitive problems (Coyle et al., 1985). Specific conditions, such as Korsakoff's syndrome and Alzheimer's disease, affect mental functions and can cause extreme memory loss. Research on animals is improving the understanding of the mechanisms of such losses. In turn, this increased understanding has led to the discovery of some drugs that show promise of counteracting the losses. Most of the knowledge about the neurotransmitters involved in these diseases has also been derived from studies of the brains and nervous systems of animals.

Primates are phylogenetically closer to humans than are other mammals. Their behavioral capabilities are in keeping with the greater development and complexity of their brains. Primates also have age-related decrements in memory function. Generally, memory impairment with advancing age first appears as a failure of immediate memory, the recall of events that have just occurred. Transmitter chemicals of the α-adrenergic class, like clonidine, were first found to improve memory performance in macaques and aged rodents. Clonidine has now also proved effective in improving the memory of patients with Korsakoff's syndrome. Those findings suggest a new approach to the treatment of patients with memory disorders, and they have provided a new option for clinical trials with patients suffering from Alzheimer's disease (Arnsten and Goldman-Rakic, 1985).

Pain

Pain is a common symptom of disease in humans and animals. It is important that medical science develop more effective methods of pain management than are now available. Much pharmacological research has focused on the production of drugs with potent analgesic properties, and much research on pain—particularly that concerned with analgesics, acupuncture efficacy, hypnosis, and so on—has been carried out on human subjects for over a century. Research using animals is necessary, however, if unsolved problems are to be adequately addressed.

Although many experiments that study pain must involve pain for the animal, researchers have developed methods that are as humane as possible within the context of the experiment. For example, the slightest reflex movement of the tail of a rat or mouse is objective evidence that a noxious stimulus applied to the skin of the tail has attained threshold intensity. Reflex behavior, such as the tail-flick, is a useful index of the comparative effectiveness of analgesics, as well as of the effects of manipulating chemical messengers in the central pain pathways (Willis, 1985).

The understanding of intrinsic brain mechanisms of pain and its modification will require the use of modern techniques for cell marking and pathway tracing, immunocytochemical and microphysiological methods, and sophisticated behavioral studies. Paradoxically, many investigations of pain can be explored in anesthetized animals. Thanks to psychophysical studies in humans that were replicated in animals, neuroscientists have been able to trace the nerve fibers from skin, muscle, and internal organs that are specific carriers of "pain signals." With such a powerful handle on the input end of the pain system, the passage and transformation of pain signals can be explored in complex neuronal organizations in anesthetized animals. It is also possible to study the central systems that control the passage of pain signals to higher levels of the central nervous system. Finally, isolation and identification of the transmitters, structure, and other components of the neurochemical machinery of the brain involved in pain perception and its modification can be elucidated (Willis, 1985).

Increasing recognition that behavioral factors play a significant role in many current health problems—for example, drugs and alcohol abuse, eating disorders, effects of stress, cardiovascular disease, and mental and psychiatric ailments—has led to the development of animal models for experimental and biological analysis as part of the emerging field of behavioral medicine (Hamburg et al., 1982).

OTHER BENEFITS FOR HUMANS

The preceding examples provide a sampling of the contributions that research using animals has made to the improvement of human health and the acquisition of knowledge. Many others could be cited—for example, the development of medicinals such as the sulfonamides (Hubbard, 1976); the development of life-support systems for premature infants (Coalson et al., 1982; deLemos et al., 1985; Escobedo et al., 1982); and the increase in understanding of learning (Miller, 1985; Pavlov, 1927; Skinner, 1938; Thorndike, 1898), nonlinguistic communication (Gardner and Gardner, 1969; Romski et al., 1984), drug abuse (Deneau et al., 1969; National Institute of Drug Abuse, 1984; Seevers, 1968), and nervous system regeneration. Many examples of such benefits are also chronicled in publications such as those by Gay (1986), Leader and Stark (1987), and Paton (1984).

BENEFITS FOR ANIMALS

One might have the impression that animal research is conducted only with the aim of alleviating human suffering. The conduct of extensive research in veterinary schools and other institutions indicates that that is not the case. Most research on domestic farm animals is undertaken to increase the productivity and quality of animal products. Research is also undertaken to reduce the suffering and increase the overall well-being of animals, particularly companion animals. Examples include current research on Potomac fever in horses, the development of ivermectin to eradicate parasitic diseases in a variety of animals, and the development of vaccines for feline leukemia virus and canine parvovirus.

Research aimed at human illnesses has also had immeasurable benefits for animals. A host of immunizations and antibiotics have proven applicable to the therapy of animal diseases (Paton, 1984). Kidney transplantation, cardiovascular treatments, chemotherapeutics, and narcotics are widely applicable, as are the insights gained from genetic research (Gorman, 1988).

One example of the benefits of biomedical research for animals can be found in the propagation of endangered species. The ability to transfer embryos, eliminate parasitism, treat illnesses, and use anesthetic advances has improved the health and survival of many species. The knowledge gained from genetic studies has allowed appropriate management of species that are endangered or have disappeared in the wild. For example, the ability to identify the sex of birds has been essential in the management of the whooping crane and the California condor. Research into obstacles to successful breeding in captivity has markedly reduced the need for importation of many species, especially monkeys. For example, among nonhuman primate species used in research, there were 7,908 births in 1984 in the United States, compared with 2,198 in 1973 (Johnsen and Whitehair, 1986).

SUMMARY

Animal research has resulted in enormous benefits for humans and animals. The searching and systematic methods of scientific inquiry have greatly reduced the incidence of human disease and have substantially increased life expectancy. Those results have come largely through experimental methods based in part on the use of animals, as illustrated by the many examples cited in this chapter.

At the same time, much obviously remains to be learned. Further studies in such areas as cancer, heart disease, diabetes, AIDS, dementias, and the development of vaccines and chemotherapeutic agents will continue to require the use of animals.

4
Alternative Methods in Biomedical and Behavioral Research

In recent years a great deal of attention has been focused on the use of alternative methods in animal experimentation (National Institutes of Health, 1981; National Research Council, 1977; Office of Technology Assessment, 1986). This interest has arisen in part because of a concern for the animals' welfare and the increasing costs of animal purchase and care. However, the term "alternative" has caused a great deal of confusion, because it implies that there are replacements for animals in many experimental situations. In reality, there are few situations in which computer simulations, in vitro techniques, or other methods are suitable replacements for animals.

By expanding what is considered to be an alternative to include reductions in the use of animals and refinements in experimental protocols that lessen the pain of the animals involved, the possibility of using alternatives increases. In addition, the replacement of one animal species with another, particularly if the substituted species is nonmammalian, can be considered another alternative method. In the following chapter we will apply this broader definition of alternatives, one that arises directly from the concepts of Russell and Burch (1959) and that was used by the Office of Technology Assessment (1986) in its report ***Alternatives to Animal Use in Research, Testing, and Education.***

Scientists searching for alternative methods have asked: How can nonmammalian organisms, in vitro techniques, and nonbiological approaches be used? To answer that question, one must first determine how alternative approaches can provide results that are relevant to humans, and how knowledge of more complex forms (organisms, organs, and tissues) can be inferred from research on cells and molecules.

Many similarities in structure and function among mammals make them obvious candidates for research applicable to humans. Rodents—rats, mice, guinea pigs, and hamsters—have been used because their small size makes them suitable for laboratory experiments and because they can be bred readily in captivity. Less well known has been the ongoing and successful use of lower vertebrates, invertebrates, and microorganisms in biomedical research.

A variety of organisms have been used in achieving the progress that has been made in biomedical research during this century (National Research Council, 1985b). For instance, of the 135 recipients of the Nobel Prize in physiology or medicine from 1901 to 1984, the majority of organisms used in their prize-winning work were mammals. One-third of the recipients were cited though for work that involved no warm-blooded vertebrates. An additional 17 were cited for work involving only humans. Twenty-five of the Nobel Prize winners based their work on a combination of different experimental subjects, including vertebrates, invertebrates, and cultures. Even higher plants have been used as sources of model systems. A survey of this series of awards is but one indication of the contributions made by a variety of organisms to biomedical research.

RELATIONSHIPS AMONG LIFE FORMS

The principle on which the search for alternatives to mammals in research depends is that of "unity in diversity" (National Research Council, 1985b). Diversity is seen in the millions of species that have existed or now exist, each of which has characteristics sufficiently different to enable them to be distinguished from one another. Unity is seen in common anatomical features and in the universality of the cell theory. For example, the development of all vertebrate embryos follows a program of blastula formation and development of ectoderm, mesoderm, and endoderm—a program that is characteristic of most invertebrates as well.

Unity is also seen in the universal scheme of intermediary metabolism, which can be displayed on a chart and involves the relatively small number of approximately 1,000 intermediates. The intermediary metabolism of any species is a subset of the universal scheme, which therefore has as much generality for biochemistry as the periodic table does for chemistry (Sallach, 1972). Furthermore, discoveries in molecular biology have demonstrated the universality of the genetic code, which applies from the simplest virus all the way to humans.

The feature that makes it possible to substitute different species and other systems—such as cell and tissue cultures, single cells, and nonliving systems—is the presence in biology of generalizations that apply quite broadly.

ANIMAL MODELS

In the last decade, knowledge and use of models and the capability of computer systems have expanded. For instance, a committee of the National Research Council (1985b) has recommended in its report *Models for Biomedical Research: A New Perspective* that NIH support those proposals aimed at the development of model systems for specific fields of research. The committee also recommended that NIH regard proposals for the study of invertebrates, lower vertebrates, microorganisms, cell and tissue culture systems, and mathematical models as having the same potential relevance to biomedical research as proposals for work on mammalian models. In addition, it recommended that NIH strive to make information on model systems readily available to the research community.

The importance of both mammalian and nonmammalian animal models to basic research and to the understanding of human disease is illustrated in the story of how researchers came to understand myasthenia gravis (Morowitz, 1986). Myasthenia gravis is a disorder characterized by muscular weakness that can proceed to complete paralysis of some muscle groups. Perhaps the first link forged in the chain of knowledge concerning the cause of myasthenia gravis was Bernard's research with frogs on the mode of action of curare, which causes paralysis of muscles. Later, it was demonstrated that muscles of myasthenic patients, when stimulated through their nerves, fail to respond, as though they have been poisoned with curare.

Fifty years ago, Loewi, Dale, Feldberg, and Vogt established in frogs and other laboratory animals that transmission of a signal from nerve to muscle was effected by the release of a chemical, acetylcholine, from the nerve ending (McGrew, 1985). The concept soon evolved that acetylcholine interacted with receptor molecules on muscle where the nerve terminated. Curare blocked the action of acetylcholine and so decreased its effectiveness.

Subsequently, two chemists in Taiwan isolated a powerful toxin from snake venom that paralyzed animals by binding to and blocking (inactivating) the receptor for acetylcholine on the surface of muscle cells. Having a chemical that could tightly bind to the acetylcholine receptor, other investigators used the toxin to obtain large quantities of acetylcholine receptor molecules from the electric eel *Torpedo*, whose electricity-producing organ contains large quantities of acetylcholine and acetylcholine receptors. The acetylcholine receptor, a protein, is now under intense investigation to determine its amino acid structure and its mode of response to acetylcholine (Kandel and Schwartz, 1985).

Receptor proteins can also produce antibodies. Indeed, in an attempt to make such antibodies, scientists injected the acetylcholine receptor protein into rabbits. Unexpectedly, the rabbits developed a complete clinical picture of myasthenia gravis, which led to the recognition that myasthenia gravis is an autoimmune disease. In fact, it is now one of the most completely understood autoimmune diseases. For some reason, the body produces antibodies that specifically bind to and decrease the functional activity of acetylcholine receptors.

The search for an understanding of myasthenia gravis in humans has involved frog muscles, rodent neuromuscular synapses, snake toxin, electric eel receptors, and rabbit antibodies. Additional research will be needed before a full cure to the disease is found—research that will probably continue to use nonmammalian models.

The preceding example illustrates how both mammalian and nonmammalian models can be used in the discovery of causes and treatments of human diseases. It also demonstrates that biomedical research requires the use of animals, whether they be frogs, rabbits, snakes, or electric eels.

ALTERNATIVES TO MAMMALS

As discussed in Chapter 2, rodents (rats, mice, guinea pigs, and hamsters) and lagomorphs (rabbits) are the mammals used most often in research. The use of some kinds of mammals is limited by their size, cost, and availability and by the emotional attachment of humans to them. Depending on the type of research in question, mammals sometimes can be replaced with nonmammalian vertebrates, invertebrates, microorganisms, cell and tissue cultures, and nonbiological systems, as discussed below. The necessary verification of experimental results still requires the use of some mammals in establishing a model system.

Nonmammalian Vertebrates

Nonmammalian vertebrates—fish, amphibians, reptiles, and birds—are rather closely related to mammals. Most of the basic properties of chemical transmission in nerve cells were learned by studying the frog neuromuscular junction (the synapse between nerve and skeletal muscle). Many similarities in embryonic development are present throughout the vertebrate class.

Invertebrates

Among the invertebrates, the largest number of species are insects. A great deal of research has been conducted on insects, and much of it has provided fundamental insights into the processes of all living things. For instance, research on the eye pigmentation of *Drosophila* led to the hypothesis that each gene controls a single enzyme—a concept that has proved fundamental to modern molecular biology (Ephrussi, 1942). Other invertebrates have also been studied; for example, research on the squid giant axon provided the basis for the concept of the ionic nature of the electrical action potential in nerve transmission (Hodgkin and Huxley, 1952).

Microorganisms

Microorganisms are acceptable as models in metabolism, genetics, and biochemistry, and they can sometimes serve as models of more complex systems. For instance, insights into the fundamental mechanisms of gene expression are applicable to the study of the normal and pathological development of human embryos. Investigators have also shown that yeast has receptors for estrogen that appear identical in affinity with those of the rat uterus (National Research Council, 1985b).

Cell and Tissue Cultures

Cell and tissue culture systems are used in basic research, in applied research on such subjects as cancer chemotherapy, and in testing of potentially toxic substances. They are relatively easy to manipulate, and living cells can be observed with a light microscope while various components of the system are changed. For instance, one can observe the beating of cultured heart cells and note the effects of adding various chemicals to the culture medium.

Human Tissues

The use of human tissues removed at surgery or at autopsy is another alternative to the use of live animals in research. Such material is available at most research centers and is similar to tissues that are the targets of the research.

One example of a human tissue used in research is the pituitary gland. Hormones from pituitary glands have been characterized, and in the past growth hormone was extracted and used to treat children with growth hormone deficiency. More recently, the latter practice has been discontinued because a few recipients have contracted Creutzfeld-Jakob disease, apparently as a result of infection with a slow virus contained in pituitaries of infected persons (Gibbs et al., 1985). A bioengineered growth hormone produced by the bacterium *Escherichia coli* that has recently become available eliminates the possibility of contracting the disease.

Human placental tissue is also used. The endothelial cells harvested from umbilical cords are used for tissue culture; the membranes are studied to further the understanding of human labor processes and have displaced, to a degree, experiments in sheep; and the placenta proper is used to study laminin and other basement membrane proteins (Charpin et al., 1985).

Various other tissues are collected at autopsy for an array of research uses. For instance, breast tissue is used to investigate the pathogenesis of cancer, and other organs are used in cardiovascular and pulmonary research.

In Vitro Systems and Mathematical Models

In vitro approaches are appropriate for some research in biology. For instance, much of the study of intermediary metabolism uses synthesized biochemicals in a manner similar to that of any chemistry experiment. Studies of reaction rates and the role of catalysts are typical examples.

Mathematical models can supplement experimental work or occasionally replace it. Such models can increase the effectiveness of experiments by defining variables and checking theories, thus making experiments on biological systems more effective and economical.

Refinements

The preceding discussion has focused primarily on means of reducing the number of animals used or replacing mammals with other organisms or with in vitro or mathematical models. The third alternative is to refine experimental techniques to lessen the discomfort of the animals involved. Such refinements of protocol constantly occur, as researchers expand the range and uses of anesthetics, improve or eliminate restraining devices, minimize invasive techniques, or use noninvasive methods to obtain the required results. Each success in this area minimizes discomfort to experimental animals.

ALTERNATIVE METHODS IN TESTING

In recent years more attention has been focused on finding alternatives to the use of animals in testing. Several centers are looking for alternatives to such tests as the Draize test and the LD_{50} test. Many of the tests are described in Chapter 2. This section examines efforts to reduce the numbers of animals used, replace a mammal or vertebrate with a lower organism, or refine procedures to reduce the pain and suffering experienced by animals in these tests. A listing of alternative methods follows.

- *Acute toxicity tests* Alternatives to the LD_{50} test have been developed that use far fewer animals, with more attention being paid to morbidity and symptoms than to a statistical estimate of the median lethal dose (Rowan and Goldberg, 1985).
- *Eye and skin irritation tests* Modifications to the Draize test have been developed that use smaller or more dilute doses of irritant substances and result in less trauma and distress to the test animal (Gloxhuber, 1985). Corneal and other cell cultures might also prove to be replacements for the Draize test. In one test system being developed, fertilized chicken eggs are used to evaluate both skin and eye irritancy; the irritant is applied to the chorioallantoic membrane, which surrounds the developing embryo (Luepke, 1985). Research and development is in progress with the hope that one or more of these methods can be validated.

- *Repeated-dose chronic toxicity tests* Cell cultures may be useful adjuncts to animals for specific target organs and tissues and thus may prove useful in routine screening tests (Office of Technology Assessment, 1986).
- *Carcinogenicity tests* A battery of short-term tests using mostly bacterial, yeast, cell-culture, and in vitro assays has been proposed as a predictor of the carcinogenicity of new and existing chemicals (Lave and Omenn, 1986). A recent study, however, has shown that four such tests have a concordance of only 60 percent with rodent carcinogenicity tests (Tennant et al., 1987).
- *Developmental and reproductive toxicity* The chick embryo has been investigated as a possible screen for teratogens, and fish and amphibian embryos, as well as other systems, might also prove useful. No single in vitro system can yet replace animal testing.
- *Neurotoxicity tests* Invertebrates can be used for some screening purposes because their nervous systems are sufficiently complex and biochemically related to the human nervous system. The developing chick embryo is being used to measure the effects of certain drugs, because the activity of the embryo can easily be observed and recorded (Norton, 1981).
- *Mutagenicity tests* The most commonly used test for mutagenicity is the *Salmonella*/microsome, or Ames test, which uses microorganisms and animal tissues (Ames et al., 1975). However, whole-animal use is also needed in certain instances—for example, to test hereditability of mutations (Office of Technology Assessment, 1986).
- *Biological screening tests* Cell and fixed-enzyme systems are used for screening whenever possible.

If suitable alternative methods can be found for these tests, reductions in the number of animals used are possible. Testing for pregnancy once relied exclusively on live animals—mostly rabbits but also mice and frogs. It is now conducted with such procedures as agglutination, radioimmunoassay, and enzyme immunoassay.

The search for alternatives to the use of animals in testing is growing rapidly. *Tox-Tips*, a journal published since 1976 by the National Library of Medicine in Bethesda, Maryland, is designed specifically to prevent duplication of toxicity-testing programs and to provide citations of tests that minimize the use of live animals. For example, the June 1986 volume included references to "Hen's Egg Chorioallantoic Membrane Test for Irritation Potential" (Luepke, 1985), "Biopharmaceutical Test of Ocular Irritation in the Mouse"

(Etter and Wildhaber, 1985), "Testing for the Toxicity of Chemicals with *Tetrahymena pyriformis*" (Yoshioka et al., 1985), and "An Approach to the Detection of Environmental Tumor Promoters by a Short-Term Cultured-Cell Assay" (Moule, 1984).

Despite these and other efforts, success in eliminating the use of animals from tests has been minimal. This lack of success is due both to the paucity of suitable alternatives and to regulations that require the use of specific animal tests. Although better models may become available that eliminate the use of animals, for the immediate future more realistic goals are reductions in the number of animals used, replacements of mammals with nonmammalian systems, and experimental refinements that lead to a reduction in the pain and discomfort of the animals being tested.

In any discussion on alternatives, it should be noted that most research using cells, tissue cultures, or nonmammalian systems is conducted not as an alternative to the use of mammals but because the system best answers the question under study. Thus, a physiologist may conduct experiments on an insect not as an alternative to mammals, but because there are questions to be answered about insects. For the same reason, when the molecular or cell biologist uses in vitro systems it is because they are the best ones available to answer his or her questions.

SUMMARY

Although the search for alternatives to the use of animals, particularly mammals, remains a valid goal of researchers, there is no chance of replacing all animals in research and testing in the foreseeable future. Nevertheless, some successes have occurred in developing nonmammalian models, in reducing the numbers of animals used, and in refining experimental protocols to reduce the animals' pain. Such research should continue, but any hope for sudden success must be tempered by the realization that progress in this area has been slow.

5
Regulatory Issues

Society has established regulations requiring that the welfare of animals used in experimentation be ensured. In addition to the self-imposed constraints of the researchers, external regulations occur at many levels and in many forms. These external regulations include formal, legal requirements—federal, state, or local legislative and regulatory controls—and research funding or journal publication contingent on adherence to specific policies.

APPROACHES TO REGULATION

Regulation may center on the nature of the problem under investigation, the procedures used, the setting in which the research is conducted, the species of animals used, or the qualifications and training of the investigator(s). Although some individuals feel that regulations should be applied according to the value society places on the expected results of the research, it would be inadvisable to develop regulations on such a basis. Indeed, applying regulations on this basis misses one of the essential aspects of science that is so crucial to its success. Almost all major results are the achievements of communities of investigators, who share their results in the scientific literature, build on each other's output, criticize and evaluate each other's work, and finally arrive at collective judgments of the validity of hypotheses based on data. It is not possible to know which results will become parts of a final structure and which are scaffolding that will make the final structure possible. Most scientific work turns out to be scaffolding that cannot readily be discerned in the final structure. The concept that scientists should use animals only for studies that lead to therapeutically useful results is therefore inconsistent with the foundations of science.

It is important to emphasize that there is no way to predict in advance what will and will not be productive research. What is important to recognize is that at the time it is undertaken, competent research has the potential to be productive.

For example, in the early twentieth century, Ehrlich used thousands of mice in 605 unsuccessful attempts to develop a chemotherapeutic agent for trypanosome diseases. Compound 606, salvarsan, proved effective against both trypanosomes and spirochetes. There was no way to predict that his sacrifice of mice would lead to the main therapy used in syphilis treatment for 30 years. The therapy came about because of Ehrlich's collaboration with organic chemists and his having read Schaudinn's papers on syphilis (De Kruif, 1926). To make animal use contingent on an assumed direct route to a predicted therapeutic payoff is to misunderstand the nature of the scientific enterprise.

Regulations applied most broadly might cover animal species with well-developed nervous systems or might be narrowed to cover vertebrates or warm-blooded animals. Traditionally, decisions on this question have been based on the animal's evolutionary relationship to humans and its capacity to suffer pain. For example, more stringent requirements might be imposed on research using nonhuman primates, due to their close biological relationship to humans.

These regulations might also be tempered by societal demands. Domestic animals, including dogs, cats, and horses, might be afforded a greater degree of consideration due to the relationship that they share with humans. This issue arises in the use of pound animals in research and attracts considerable public attention (see Chapter 6).

Many people consider the amount of pain or suffering, both mental and physical, inflicted on animals to be one of the most important issues surrounding their use in research. This concern has led to recent classifications of categories of biomedical research based on considerations of pain or suffering—for example, by the New York Academy of Science (1988), the Scientists Center for Animal Welfare (Orlans, 1987), and the British government in the Animals (Scientific Procedures) Act of 1986. In general, the more painful the procedure, the stricter the regulation proposed. These classifications emphasize that investigators should attempt to reduce pain whenever possible and to explore alternatives to painful procedures.

The issue has also been addressed in a regulatory proposal of the U.S. Department of Agriculture published in the *Federal Register* of March 31, 1987 (pp. 10313–10314) and in a recent American Veterinary Medical Association colloquium on the assessment of pain and distress in animals (Colloquium, 1987).

FEDERAL REGULATIONS

Congressional interest in the humane treatment of animals is not new; legislation was first passed in 1873. That law, the Twenty-Eight Hour Law, limited the number of consecutive hours that livestock could be confined for rail transport. In 1958, such concerns also led to passage of the Humane Slaughter Act (P.L. 85-765), which stipulated that animals must be slaughtered by humane means.

The Laboratory Animal Welfare Act of 1966 (P.L. 89-544), which was passed primarily to protect pet owners, addressed mounting public concern over the theft and subsequent sale of pets to research facilities. As stated by the original act,

> ...to protect the owners of dogs and cats, from the theft of such pets, to prevent the sale or use of dogs and cats which have been stolen, and to insure that certain animals intended for use in research facilities are provided humane care and treatment, it is essential to regulate the transportation, purchase, sale, housing, care, handling, and treatment of such animals by persons or organizations engaged in using them for research or experimental purposes or in transporting, buying or selling them for such use.

The Animal Welfare Act of 1970 (P.L. 91-579) broadened references in the Act from dogs and cats to animals more generally defined. In addition to research, it also added exhibition purposes and use as pets as covered activities. In the Animal Welfare Act Amendments of 1976 (P.L. 94-279), the preamble was reworded to reflect increased emphasis on humane care and treatment for research, exhibition, or pet animals.

Although the 1966 law specified six groups of animals as covered—dogs, cats, nonhuman primates, rabbits, hamsters, and guinea pigs—record-keeping requirements applied only to dogs and cats. The law required identification of dogs and cats kept on the premises of animal dealers and laboratories, and it required dealers to be licensed and laboratories to be registered. Only laboratories that used dogs or cats and either received federal funds or purchased animals in commerce were required to register. The 1970 amendments extended the authority of the secretary of agriculture to protect all species of warm-blooded animals in laboratories, as well as in the wholesale pet and exhibition trades. To date, however, the secretary of agriculture has not extended coverage under the Act to rats, mice, birds, and farm animals used in biomedical research, although rats and mice account for about 85 percent of the animals used in research, testing, and education.

In 1966 the secretary of agriculture was instructed by Congress to promulgate minimum standards for housing, feeding, watering, sanitation, ventilation, and shelter from weather and temperature extremes. Adequate veterinary care was also mandated; however, the secretary was explicitly prohibited from prescribing standards for the handling, care, or treatment of animals during actual research or experimentation by a research facility as determined by such research facility.

Appropriate use of anesthetic, analgesic, or tranquilizing drugs was added to modify the concept of adequate veterinary care in 1970. More significantly, the blanket proviso protecting the conduct of research was altered to require the research facility to demonstrate, at least annually, that professionally acceptable standards governing the care, treatment, and use of animals were being followed during actual research or experimentation.

The Animal Welfare Act Amendments of 1976 focused on transportation of animals and animal fighting. Specifically, the changes broadened the law to:

- cover regulated carriers, intermediate handlers, and animal brokers, requiring them to adhere to humane standards;
- protect all dogs, including dogs for hunting, security, or breeding purposes;
- restrict transportation of animals by prohibiting C.O.D. shipment unless the shipper guaranteed round-trip payment of care costs for animals not claimed at the destination and by prohibiting transportation of animals less than a certain age;
- require a health certificate signed by a licensed veterinarian to accompany animals transported in commerce;
- strengthen civil penalties for violations of the humane standards;
- make criminal the promotion of, sponsorship of, or partici-

pation in fights between mammals or cocks, except where explicitly permitted by state law; and

- require federal agencies to comply with the standards and other requirements of the Act.

Legislation Passed in 1985: The Health Research Extension Act and the Food Security Act

Two laws enacted in 1985 contain provisions that apply to the regulation of animals used in research. The first, the Health Research Extension Act of 1985 (P.L. 99-158), popularly called the NIH Reauthorization Act, applies to all research funded by the PHS. This legislation served to transform into law many of the provisions contained in the *Public Health Service Policy on Humane Care and Use of Laboratory Animals by Awardee Institutions*. (This publication was reentitled the *PHS Policy on Humane Care and Use of Laboratory Animals* in the September 1986 revision.) Major points of the legislation require that:

- research facilities establish institutional animal care and use committees including at least one veterinarian and one individual not affiliated with the institution;
- animal care committees review the care and treatment of animals at least semiannually;
- institutions make available training that includes information on the humane practice of animal care and use and the concept, availability, and use of research or testing methods that minimize animal distress and the number of animals used; and
- applicants for NIH funds file assurances with NIH certifying that the investigator and the institution adhere to the NIH guidelines.

The law also requires that every applicant for NIH funds include a justification for the use of animals in that research.

The Improved Standards for Laboratory Animals Act was passed in December 1985. These amendments to the Animal Welfare Act were incorporated into the omnibus farm bill reauthorization, the Food Security Act of 1985 (P.L. 99-198). In addition to provisions directly affecting the care and use of animals in research, these amendments direct that an information service at the National Agricultural Library be established to disseminate information that will reduce the unintended duplication of animal experiments, to provide information on alternatives to laboratory animals, and to provide information on humane practices for scientists and other research personnel. The concern over minimizing unnecessary duplication is echoed in the findings that preface the legislation:

1. the use of animals is instrumental in certain research and education for advancing knowledge of cures and treatment for diseases and injuries which afflict both humans and animals;
2. methods of testing that do not use animals are being and continue to be developed which are faster, less expensive, and more accurate than traditional animal experiments for some purposes, and further opportunities exist for the development of these methods of testing;
3. measures which eliminate or minimize the unnecessary duplication of experiments on animals can result in more productive use of federal funds; and
4. measures which help meet the public concern for laboratory animal care and treatment are important in assuring that research will continue to progress.

These statements highlight the shift in major emphasis, from preventing the stealing of pets to protecting laboratory animals, that has occurred through subsequent amendments to the Animal Welfare Act.

The law requires that each registered facility appoint an institutional animal committee that includes a veterinarian and a person not affiliated with the institution to represent general community interests in the proper care and treatment of animals. The provisions for committees, present in both 1985 laws, bring the overwhelming majority of experimental animal users in the United States under the oversight of a structured, local review committee (Office of Technology Assessment, 1986).

The Improved Standards for Laboratory Animals Act provides for some specific standards. Institutional animal committees are required to inspect animal study areas twice each year and report any deficiencies to the institution for correction. If the institution does not take appropriate action, the U.S. Department of Agriculture and any funding agencies involved must be notified. As a result, grants or contracts may be suspended or revoked. The committee is also responsible for reviewing practices involving pain to animals.

Investigators are required to consider alternatives to animal use and to consult with a veterinarian before beginning any experiment that could cause pain. The standards issued by the secretary of agriculture will include provisions regarding exercise for dogs, environments adequate to promote the psychological well-being of primates, presurgical and postsurgical care, the use of pain-relieving drugs, euthanasia, prohibition of the use of paralytics without anesthesia, and prohibition of the use of an animal for more than one major surgical procedure. Any exceptions to the standards set forth in the law or in the regulations promulgated under the law must be specified in the research protocol and justified in a report filed with the committee.

The 1985 amendments mark the first time that the practice of animal experimentation itself has been opened to public scrutiny through the institutional animal committee. Although the Improved Standards for Laboratory Animals Act contains the proviso that nothing shall be construed as authorizing the secretary to promulgate rules, regulations, or orders with regard to the performance of actual research or experimentation by a research facility as determined by that research facility, the new law does require the facility to demonstrate that professionally acceptable standards governing animal care, treatment, and use, including the use of anesthetics, analgesics, and tranquilizers, are being followed during experimentation.

GOVERNMENT POLICY STATEMENTS

Guide for the Care and Use of Laboratory Animals

The *Guide* was developed for NIH by the National Research Council's Institute of Laboratory Animal Resources (National Research Council, 1985a). The PHS has a long-standing policy of requiring adherence to this document's guidelines by its intramural researchers and by extramural grantees and contractors that use living warm-blooded vertebrates in research and testing. The *Guide* provides a framework for the animal care and use policies of many federal agencies, nonfederal government agencies, and private organizations. For instance, the American Association for Accreditation of Laboratory Animal Care (AAALAC) uses the tenets of the *Guide* in evaluating the animal facilities of institutions that are seeking accreditation.

The *Guide* was first developed in 1963, before passage of the Laboratory Animal Welfare Act. It has been revised five times, most recently in 1985. It is considered a living document, subject to modification in the light of changing conditions and new information. The guidelines are based on established scientific principles, expert opinion, and experience with methods and practices consistent with humane, high-quality animal care. The *Guide* is written in general terms, so that it can be adapted to suit the needs of the widely varying scientific institutions that use live vertebrates. It is important to note that application of professional judgment is an essential component of the *Guide*.

Public Health Service Policy on Humane Care and Use of Laboratory Animals

The PHS policy on humane care and use of laboratory animals, which was revised in 1985 and 1986, requires institutions to establish and maintain proper measures to ensure the appropriate care and use of all animals involved in research, research training, and biological testing activities conducted or supported by the PHS. The PHS policy (Public Health Service, 1985) requires compliance with the Animal Welfare Act and its implementing regulations as well as with the current edition of the *Guide for the Care and Use of Laboratory Animals*. The 1986 revision of the policy incorporated the changes in the Public Health Service Act mandated by the Health Research Extension Act of 1985 (discussed earlier). It now specifies criteria and procedures for providing institutional "animal welfare assurance" to the NIH Office for Protection from Research Risks, which administers the assurance program. The Health Research Extension Act places the force of law behind much of the PHS policy.

The PHS policy mandates that an institutional animal care and use committee (IACUC) be appointed by the chief executive officer of each institution. Each IACUC must have at least five members and include at least one doctor of veterinary medicine, one practicing scientist experienced in research involving animals, one member whose primary concerns are nonscientific, and one member who is not affiliated with the institution in any way (other than as a member of the IACUC) and is not a member of the immediate family of a person who is affiliated with the institution. Among the committee's functions are reviewing the animal care and use program, inspecting animal facilities at least twice a year, preparing reports, and reviewing specific activities and concerns. The IACUC is also empowered to suspend animal-related activities that are not in compliance with the requirements of the policy.

The policy specifies procedures for reviewing, applying for, and reporting PHS-conducted or PHS-supported research involving the care and use of laboratory animals. Each institution must assure the PHS that it is accredited by AAALAC or another accrediting body recognized by the PHS (in addition to being evaluated by the IACUC and reevaluated periodically by the accrediting body) or is evaluated only by its IACUC and reevaluated at least once every six months. PHS staff and advisors can also review each awardee institution (which may include site visits) at any time to assess the adequacy or accuracy of the institution's compliance or expressed compliance with the policy.

The PHS policy endorses and is intended to implement and supplement the "U.S. Government Principles for the Utilization and Care of Vertebrate Animals Used in Testing, Research and Training" developed by the Interagency Research Animal Committee (IRAC). This one-page document, containing nine numbered principles, incorporates by reference the *Guide for the Care and Use of Laboratory Animals*. IRAC, which was established by the federal government in 1983, serves as a focal point for discussions by federal agencies of issues involving animal species needed for biomedical research and testing. Its primary concerns are the conservation, use, care, and welfare of research animals. Its responsibilities include information exchange, program coordination, and contribution to policy development.

Good Laboratory Practices Regulations

The Good Laboratory Practices (GLP) regulations of the Food and Drug Administration (effective as of June 1979) and Environmental Protection Agency (effective as of September 1985) are aimed primarily at ensuring efficiency and accuracy in testing procedures and do not address animal welfare directly. However, in requiring sanitation and proper maintenance of test animals, they address and influence the well-being of animals.

Facilities, including those for animals, are covered in the GLP regulations that require rooms to allow for separation of species or test systems, isolation of individual projects, quarantine of animals, and routine and specialized housing of animals. Ancillary space is required for food, bedding, diagnostic purposes, and veterinary medical treatment and control. Procedures for animal care and record retention are also specified.

A 1984 memorandum of understanding among NIH, the U.S. Department of Agriculture, and the Food and Drug Administration provides for the sharing of information based on observations made by the several agencies in site visits or inspections.

STATE REGULATIONS

General anticruelty laws were passed in every state between 1828 and 1913. However, 23 states specifically exclude animal experiments conducted in scientific institutions from the provisions of the anticruelty statutes.

State laws were passed in the years after World War II to require release of impounded dogs to research institutions on request (this subject is discussed in greater detail in Chapter 6). Most states have since repealed those laws, although 5 states and the District of Columbia still require release on request. Twelve states prohibit release of impounded animals for research, and Massachusetts prohibits experimental use of impounded dogs not only from Massachusetts but from jurisdictions outside the state. It also provides for inspections of laboratories by licensed humane officers from the Massachusetts Society for the Prevention of Cruelty to Animals and the Boston Animal Rescue League.

There is an increasing trend toward regulation of research facilities at the state level. The New York State Department of Health's Wadsworth Center for Laboratories and Research first issued regulations in 1952 under Title 5 of the Public Health Law for approval of laboratories that use living animals in research. These regulations were most recently revised in 1983. The standards follow the federal Animal Welfare Act and the *Guide for the Care and Use of Laboratory Animals*. Twenty states and the District of Columbia have state laws for the licensing of research facilities (National Association for Biomedical Research, 1987).

Instances of suffering inflicted on animals in projects for science fairs have led to the enactment of state laws that prohibit painful experiments on animals by students below college level. Such laws have been passed in California (1973), Maine (1975), Massachusetts (1979), Florida (1985), and New Hampshire (1985). In addition, voluntary guidelines have been adopted by many organizations that sponsor science fairs.

APPROACHES TO REGULATION

Granting Agency Approach to Regulation

An example of the granting agency approach to regulation is provided by the PHS. The PHS requires that each institution provide assurance to the NIH Office for Protection from Research Risks that its animal use is in compliance with the standards set by the *Guide for the Care and Use of Laboratory Animals*. This places the responsibility for compliance and for providing assurances at the local level, with the institution itself. If an institution is not in compliance, it is ineligible to receive funds for research involving animals. The risk of such punishment can be most effective, given the dependence of biomedical researchers on federal funds.

In previous years, animal welfare advocates have felt that the PHS assurance program was relatively ineffective in monitoring the use of both human and animal subjects. Recently, NIH has undertaken a series of unannounced site visits to institutions to review their animal-use programs for compliance with stated assurances. Some of these visits have led to suspension or temporary withdrawal of NIH funds and permission for some uses of laboratory animals at several major institutions.

The recent inspections by NIH and the halting of research at major institutions have alerted the biomedical community to the seriousness of NIH's intent to require adherence to standards. When a major institution is cited for deficiencies and research funds are withheld, all institutions are reminded of the necessity to follow regulations. By the same token, the recent disclosures of violations at several institutions have allowed some individuals to reemphasize their contention that animal research should not be conducted and that the existing national regulations are neither effective nor sufficient.

Recent statutory changes have given NIH policies, which were originally intended to serve as guidelines, the authority of regulations. One effect of this has been to give the *Guide to the Care and Use of Laboratory Animals* the force of law. This poses problems that go beyond the dilemma of reaching a balance between the demands of humane treatment toward animals and research needs. For the most part, problems arise from the fact that the *Guide* was drafted initially as a codification of "good practice" and an aid to self-regulation. As such, it has served the community of individuals using laboratory animals well, providing helpful guidelines for animal use and care in the 25 years since it was first published. The *Guide* was not intended to have the force of law for setting minimum standards. It has always been intended to be used with professional judgment.

Despite the fact that the contents of the present *Guide* are not substantially different from recent versions, it is now being used to set minimum standards for inspection and may in some respects be too rigidly interpreted—for example, its requirements for cage sizes and multiple separate areas and rooms for performing aseptic surgery. With the new role of the *Guide*, the drafting of the rules and their interpretation need reasoned discussion and clarification. Greater efforts also should be made to publicize NIH decisions on interpretation of the *Guide*, so that the research community can take advantage of the common-law method of learning from experience and decisions.

Self-Regulation

There are many laws and regulations covering the use of animals in research, testing, and education. However, a great deal of activity involving laboratory animals entails self-regulation. This is true in many areas of research only partly controlled by federal regulation, including the use of human subjects and biosafety, and results from the desire to encourage free and creative inquiry within a framework of regulation. For example, the Animal Welfare Act and the Health Research Extension Act cover the care and treatment of animals in laboratory animal facilities, but neither affect, except by inference, the design of research protocols that involve animals. By law, animal care and the minimization of pain and distress are the primary objectives of institutional review. As a practical matter, however, it is difficult for committees to separate animal welfare objectives from protocol requirements and scientific content during the review process. Nevertheless, this system is intended to ensure that the use of animals in experimentation will remain the responsibility of local institutional review committees and individual investigators.

The protocol review system that now exists has been a subject of much debate. In some instances, the need to receive a preliminary approval of a research protocol can be disruptive to the researcher because efficiency may be reduced, costs may increase, and progress may be slowed. On the other hand, protocol review can help the researcher when it provides an opportunity for his or her peers to offer advice and assistance. This advice may result in a better-planned experiment that not only improves animal care and minimizes pain but may also lead to more instructive results.

Self-regulation appeals to many institutions and individual investigators because it can allow for gradual change in response to societal and peer pressures rather than forcing compliance. Regulations can only complement, not substitute for, a strong sense of stewardship by the investigator who handles animals and conducts experiments. Self-regulation is indispensable both for the proper care of animals and for the success of any research that requires well-treated animals.

Many professional societies, agencies, and research institutions have established and are establishing policies on animal experimentation; examples include the American Physiological Society, the Society for Neuroscience, and the American Psychological Association. Editors of some journals have required adherence to policies of relevant societies as a condition of publication. The policies generally require adherence to existing laws; the use of anesthetics, analgesics, and tranquilizers for interventive or painful procedures, unless they would impede the experiment; provision for bodily needs; and legal acquisition of the animals. Some voluntary codes adopted by professional societies and other groups also refer to alternatives. For example, the Society of Toxicology (1986) "encourages and supports the development of valid, scientific alternatives to current animal research testing procedures."

EFFECTS OF REGULATIONS

Effects on Training and Education

If costs continue to escalate and if regulations become more stringent, animal use will be further reduced in education and training for undergraduate, graduate, and medical students. The impact of this is unclear, as some scientists already feel that animal experimentation is of only limited value in education and accordingly have greatly reduced their use in teaching. In one large medical school, for example, the use of dogs in teaching has been reduced by 75 percent and in another no animals are used for this purpose. Other scientists and clinicians feel strongly that such a reduction will have, and has had, a major negative impact on training.

Effects on Experimental Results

Where regulations lead to even greater care to ensure that appropriate conditions exist, they can be viewed as benefiting the investigation. Animal experiments generally must be performed in a manner that permits unimpeded study of the effects of a single "challenge." For example, an experiment in which the polio virus is the challenge must not be confounded by the presence of other viruses. This fact of experimental design means that animal experiments properly are performed under hygienic conditions to avoid the presence of adventitious infectious agents or poor environmental conditions, thus ensuring the health of the animal and the reliability of the research. In some cases where lifelong observation of the animal is required, the good health of the animal is essential so that it will live its normal life span.

Effects on Direct Expense of Research

The cost of doing animal research is increasing, and nonprofit biomedical research organizations face rising costs for facilities and a reduced ability to recover totally the costs of animal care. The 1978 survey by the National Research Council's Institute of Laboratory Animal Resources (National Research Council, 1980) identified a need for $350 million at 480 institutions to bring those institutions up to the standards of the *Guide for the Care and Use of Laboratory Animals*. An additional $407 million was estimated to be needed for remodeling, additions, and space replacement through fiscal year 1988. Sixteen percent of all the biomedical research institutions surveyed needed to replace current animal facilities, 38 percent needed to remodel facilities, and 43 percent needed to provide additional space. The surveyed organizations also reported a need of $43 million for equipment renovation, replacement, or additions. At that time, a lack of space and equipment may have been the reason why 18 percent of the organizations in the confidential survey were unable to comply in toto with federal guidelines for animal care.

The ILAR survey data showed that of the $2.27 billion spent by nonprofit institutions in 1978 for biomedical and behavioral research, 35 percent ($797 million) was for projects involving animals. That decrease from the 44 percent reported in 1968 suggests that the costs of other aspects of research may be increasing, but it might also reflect a decrease in the use of particular animals in research. According to the ILAR surveys, between 1967 and 1978 the numbers of acquired research animals of all species decreased, except for swine, cattle, horses, and rodents other than mice, rats, hamsters, and guinea pigs. Factors contributing to the decrease were diminished funding, less available space, an inability to comply with federal guidelines, a shift in research methods from short- to longer-term animal studies, and increased use of alternatives. The substantial decrease (47 percent) in the use of nonhuman primates was also due at least in part to the higher costs of the animals and of their maintenance and their reduced availability caused by export restrictions.

The costs of acquiring and caring for laboratory animals have continued to increase since 1978. In large institutions, due in part to animal care regulations, animals used in research are commonly kept in centralized animal resource facilities. The cost of maintaining such a facility depends on many factors, including administration, the operation and maintenance of vehicles, animal purchasing, cage-washing, refuse disposal, feed and bedding, laboratory services, animal health care, surgical and x-ray services, research services, animal husbandry, and capital and amortization costs. Such costs are usually included in assessments of per diem charges for animals.

Although it is difficult to compare charges among institutions because procedures for determining costs and charges vary, one can get an idea of the changes that have occurred over time by looking at a single institution. For example, at one university, per diem charges for mice rose from 5.5 cents in 1978 to 14 cents in 1987—an increase of over 150 percent—while the Consumer Price Index rose by 74.1 percent (U.S. Department of Labor). Over the same period, the per diem charges increased for dogs (from $3.60 to $8.61), for monkeys (from $1.05 to $2.71), and for cats (from $1.30 to $3.20). At another university, the per diem charge for dogs rose from $2.37 in 1981 to $3.50 in 1985—a 48 percent increase. The average per diem charge for mice increased during the same period from 5 cents to 9 cents—an 80 percent increase. During the same period, the Consumer Price Index went up by 18 percent.

The cost of acquiring animals has also risen. For example, at one medical center the average purchase cost per random-source dog was $8 in 1964, $96 in 1981, and $154 in 1986. In addition, as more restrictions are placed on the use of pound animals, which have an average purchase cost of $5 to $55 and then cost perhaps an additional $100 to prepare for research use, more animals will have to be acquired from commercial breeders, who may charge between $275 and $600 for a 10-kg adult dog. Not included in this figure is the shipping cost or the additional per diem charges incurred because the health of the animal must be maintained in anticipation of the experiment.

Another indicator of the cost of compliance with PHS policy on animal care and use can be seen in the response to an initiative of the Animal Resources Program of the NIH Division of Research Resources. Announced in December 1984, the initiative provided funds to share renovation costs with the awardees on a 50–50 matching basis. The renovations were to enable institutions to comply or continue to comply with PHS policy. Ninety-nine applications were received, requesting $38,598,558; 74 were approved. However, money was available to fund only 17 of the approved projects ($7,369,000).

All the causes of the increased costs may be debated. Yet whatever the causes, the results are clear: it is expensive to use laboratory animals in biomedical and behavioral research. That expense might ultimately have great influence on the numbers and kinds of animals used in research.

Effects on the Animals

Animals are the intended beneficiaries of regulation. What needs to be assessed is to what extent animals are more humanely treated as a result of such regulations. Although difficult to measure, the following observations provide some feeling for the impact of legislation.

When the first federal law on animal experimentation, the Laboratory Animal Welfare Act of 1966, was enacted, many dog dealers ceased business because they did not wish to adhere to the required minimum standards of handling, housing, feeding, watering, sanitation, ventilation, shelter from extremes of weather and temperatures, and adequate veterinary care. While few, if any, registered research facilities stopped using animals, many discarded outdated, damaged, unsanitary, and unsanitizable cages and food and water containers and discontinued use of cages that were too small to allow animals to stand up and make normal postural adjustments. Large numbers of substandard cages were removed before the compliance deadline.

ENFORCEMENT AND ENACTMENT OF REGULATIONS AND LAWS

No discussion of regulations and their impact would be complete without some discussion of the enforcement of existing regulations and laws and the enactment of further regulations and laws. As mentioned earlier, NIH is beginning to make unannounced visits to laboratories, and the USDA has been making such visits for many years. These visits sometimes have revealed violations of the regulations. Many violations have been minor, but a few have been more serious. The more serious ones have resulted in suspension of funding and/or imposition of fines.

While it is necessary for such inspections to continue and for regulations to be enforced, it is less clear that more regulations need to be promulgated. At present, the USDA is developing regulations based on the 1985 amendments to the Animal Welfare Act. When these regulations are enacted, the scientific community should be given sufficient time to adjust to them, and their impact should be assessed before any new regulations are considered.

As undesirable as the violations have been, they do not justify break-ins at animal care facilities and laboratories, which have increased in recent years. These break-ins, for which radical animal rights groups have often claimed responsibility, have resulted in vandalism that has been costly for the institution, the individual investigator, and society, which loses the benefits of the research. Such violations of the law are not tolerable, and the offenders must be made to realize the full implications of their actions and punished accordingly.

6
Use of Pound Animals

Much of the controversy surrounding animal experimentation is related to the use of animals from pounds. This subject has become a major political issue in recent years.

SUPPLY OF POUND ANIMALS

A pound is a facility established by local ordinance in which stray, abandoned, lost, or donated animals are held—impounded—for some period, so that owners can claim lost pets or new homes can be found for the animals. A shelter is a privately established facility for such animals. In pounds and most shelters, over 90 percent of the unclaimed animals must eventually be killed. In the United States, more than 10 million dogs and cats from pounds and shelters are killed each year. The annual cost of control of stray dogs and cats in the United States is over $500 million, which includes the costs of euthanasia and disposal of these 10 million animals. Approximately 138,000 dogs and 50,000 cats are obtained from pounds and shelters each year for use in research and testing (Foundation for Biomedical Research, 1987), and most of these are used in acute, nonsurvival research under full anesthesia.

Dogs and cats obtained from pounds and shelters are described as random-source animals—the term used for any animal not bred specifically for research. Random-source animals are obtained from pounds and shelters or from USDA-licensed dealers that obtain them from pounds, shelters, farms, and other such sources. In 1983, approximately 182,000 dogs were used in research and testing in the United States, including pound and other random-source animals, as well as those bred specifically for research use (Office of Technology Assessment, 1986).

REGULATIONS

Forty-nine states permit the use of some pound animals in research. Eleven states do not allow pounds within their jurisdiction to make animals available to research facilities, but permit animals from out-of-state pounds to be purchased through USDA-licensed dealers. In Massachusetts, all use of pound animals is prohibited.

SCIENTIFIC CONSIDERATIONS

Pound animals have varied medical histories and are seen as having varied genetic backgrounds. In many experiments, the investigator may determine that this variability poses no problems or may even be of value in the experiment in that these animals provide greater diversity of genetic background and hence mimic the human situation. In other experiments it is necessary to know genetic compositions and the use of pure-bred animals is necessary. In other cases the unknown health status, physiological condition (e.g., whether they are spayed or pregnant), and age of the pound animal may introduce a chance of biological and experimental variability that could interfere not only with the results obtained but also with interpretation of the data.

NIH policy is that decisions as to the kinds and sources of animals appropriate for research be made by individual scientists and institutions (National Institutes of Health, 1987). For scientists whose research is already based on random-source animals, continued access to such animals allows them to build on extant data. It should be noted that some commercial dealers also provide randomly bred animals, but at a greater cost than that of animals from pounds.

BENEFITS

Dogs and cats obtained from pounds and shelters are used and have been used in research on a wide variety of diseases, including diabetes, cancer, arthritis, and cardiovascular ailments. For example, pound dogs were used in the development of the counter-shock treatment for restarting the human heart in patients whose hearts stop beating as a result of electric shock, heart attacks, or other causes. These animals were used to determine the most effective means for restoring the heartbeat. In addition, most current surgical methods for treating heart and kidney disease have been developed through research on dogs. Cats have been extensively used in research on the nervous system. Cats have also been used in research on visual and auditory function and may be used as a model for AIDS research, as mentioned in Chapter 3.

COST CONSIDERATIONS

Scientists seek every legitimate way to keep their costs as low as possible. They are concerned that the progress of research might be impeded if relatively inexpensive pound animals are not available. If the approximately 138,000 pound dogs used each year for scientific research were not available, there would be a need to breed and raise additional dogs to replace them. These animals would cost researchers a substantial additional amount of money every year at current levels of use.

CONCERNS FOR THE ANIMALS

Obtaining animals from commercial breeders rather than pounds not only increases expenditures but also increases the total number of animal lives lost each year. Over 10 million animals already die in pounds and shelters each year, and additional animals bred for research add to the total loss of animal life.

Some people contend that pound dogs and cats should be viewed differently from those bred specially for research purposes. Pound animals are not adjusted to the confinement of the laboratory, they assert, and may experience more stress because of the change from having been pets in homes (although many animals taken to pounds are unwanted or unsuitable as household pets). Animals that had been bred for research, having never experienced the social interaction and freedom of movement of a home environment, could be considered to be affected less by their absence. However, some breeders of dogs and cats for research include socialization and walks as part of their policy, so these distinctions are not always so clear-cut.

To avoid the concern about long-term experiments using pound animals, some individuals and humane organizations would restrict the research use of pound dogs and cats that are already scheduled for euthanasia to acute nonsurvival experiments under full anesthesia. In acute nonsurvival experiments, animals do not regain consciousness after the experiment. In chronic experiments, animals do regain consciousness. Indeed, in such experiments, not only their survival but their full recovery might be an essential part of the experiment.

7
Conclusions and Recommendations

In our society, no single ideology or theology governs people's mode of thinking. Different perspectives constitute the pluralistic base of our thought. This society is one that considers the ethics of public policy—that is, the identification of the set of values that places priorities on achieving what is considered best for the common good.

When decisions that affect the welfare of society are made, these ethical considerations are vital. Lacking a single world view tied to a specific religious or philosophical perspective, but believing in a need for a framework within which to make decisions, our society seeks to make policy decisions in science and all areas within an ethical framework.

In the United States, social decisions are imposed not by authorities but by the will of the people acting through elected representatives, whose responsibility it is to hear and consider differing voices. We live in a society based on a spirit of liberty. We must make public policy decisions within that spirit. In his address "The Spirit of Liberty," Judge Learned Hand acknowledged the difficulty of defining the precise meaning of that term (Hand, 1960). Yet he underscored an attitude of humility with which a free people must make decisions when he said

> The spirit of liberty is the spirit which is not too sure that it is right;

> the spirit of liberty is the spirit which seeks to understand the minds of other men and women; the spirit of liberty is the spirit which weighs their interests alongside its own without bias. . . .

In that context, persons with varying perspectives present their convictions on the issue of the use of animals in research. A few individuals are opposed to any use of animals in research, while the vast majority favor their appropriate and humane use.

We feel that the majority is in fact correct. Our view is not based only on an abstract desire for the advancement of science. It also arises out of a concern for those who suffer from conditions such as Alzheimer's disease, schizophrenia, manic-depressive psychoses, drug abuse, AIDS, cancer, spinal injuries, diabetes, and many other diseases that as a result of research might be prevented, alleviated, or cured.

Animal experimentation has enormously benefited humans, as well as animals, in the past and will continue to be necessary for clinical and basic research in the future. Indeed, there is no reason to believe that animal experimentation will be less productive in the future.

We are convinced that humans are morally obliged to each other to better the human condition. In cases in which research with animals is the best available method to reach that goal, animals should be used. We also believe that scientists are ethically obliged to ensure the well-being of animals in research and to minimize their pain and suffering.

RECOMMENDATIONS

The committee affirms the principle of humane care of all animals, including those used in research.

- The committee recommends that all those responsible for the use and care of animals adhere to the principle that these animals be treated humanely.

A large body of laws and regulations exists for the care and use of animals in research in the United States and internationally. In some countries, strict legislation has made it difficult to perform some research and has reduced potential contributions to human welfare through science. The committee believes it is necessary that laws and regulations be balanced to ensure the availability of animals so that research continues effectively. The present regulatory framework in the United States, if implemented properly, should meet current societal and ethical expectations and permit knowledge to continue to evolve with the appropriate balance between scientific and humane goals.

- The committee recommends that no additional laws or regulatory measures (excepting the regulations required by the Improved Standards for Laboratory Animals Act of 1985) affecting the use of animals in research be promulgated until, based on experience, a careful accounting of the effects of the application of the present body of laws, regulations, and guidelines has been made and any evidence of the need for more regulation is available.

When a number of new or revised measures are introduced in a short time, it is possible that the measures will have an untoward effect on the performance of research using animals. Yet rules, once in place, are difficult if not impossible to alter. The committee believes that there must be a mechanism for ongoing review of these measures to ensure that they not only protect animals but permit valid research to proceed.

The committee calls attention to one specific case. The *Guide for the Care and Use of Laboratory Animals* was drafted originally as a general guide for good practices in research, animal care, and use of animals and was not meant to establish minimum standards. Subsequently, the PHS adopted the *Guide* as required policy for all PHS grantees. This gave the provisions of the *Guide* authority that had not been intended when it was drafted.

- The committee recommends that there be a mechanism for ongoing review of the regulatory framework of federal agencies for animal experimentation. It is essential that research scientists who must abide by this framework be prominently involved in its assessment. Specifically, the *Guide for the Care and Use of Laboratory Animals* should be reviewed as soon as possible to determine whether revisions are necessary due to changing conditions and new information.

Institutions and individual investigators are unnecessarily burdened and confused by the differing regulations and criteria imposed by different federal agencies. The Interagency Research Animal Committee is attempting to correct this problem. Any attempt to reduce the confusion caused by the multiple authorities responsible for setting and enforcing regulations is useful:

- The committee recommends that federal standards developed by different agencies for the care and use of laboratory animals be congruent with each other.

Intentions can be effected only if the means exist to do so. Those "means" usually translate into more money. An inspection system to enhance the protection of research animals must have available to it funds to support adequate manpower and implementing structure. Yet financial support for inspection purposes has been difficult to obtain. Regulations that increase the cost of doing research—for example, the replacement of small cages with larger ones—are expensive for research institutions. Yet funds for the rehabilitation of existing facilities, for the creation of new ones, and for compliance with new rules and regulations have decreased while the need for them has increased.

It has been estimated that about $10 million annually are required for the Animal and Plant Health Inspection Service (APHIS) to operate the inspection system mandated by the Animal Welfare Act and that an estimated several hundred million dollars would be required to maintain and upgrade animal care facilities to comply with the act.

The committee recognizes the fact that there is fierce competition for funds for the support of research at a time when federal expenditures for all purposes, including research, are undergoing close scrutiny. Still, as individuals, and as a group interested in both the continuation of valid research and the humane treatment of animals used in that research, the committee is concerned that neither of these aims can be reached unless adequate financial support is provided.

The committee is also concerned that funds not be diverted from other support of research. Additional funds, not diverted funds, are required to maintain the pace of biomedical discoveries.

- The committee recommends that sufficient federal funds be appropriated for the inspections required for the enforcement of the Animal Welfare Act.
- The committee recommends that sufficient federal funds be appropriated for maintenance and improvement of animal facilities to allow individuals and institutions to conduct animal research in compliance with government policies, regulations, and laws. It is important that such funds be added to ongoing research support.

The committee focused on animal use in research rather than in testing or other areas. Large numbers of animals, however, are used in testing for the toxicity of substances found in consumer products such as food, drugs, and cosmetics. Such tests are prescribed by law and are intended to protect consumers. Considerable effort is being made to develop alternative testing methods.

The committee recognizes the need to protect the public health through testing. At the same time:

- The committee recommends that federal regulatory agencies move rapidly to accept tests—as such tests become validated—that reduce the number of vertebrates used, insofar as this does not compromise the regulatory mission of an agency and protection of the public.

In many instances a specific animal or procedure is the best or only system for performing research on a particular biological process. In some instances, though, alternative methods may be available that allow scientists to reduce the number of animals used, replace mammalian models with nonmammalian models, and refine experimental procedures to lessen any pain that animals may feel. The committee has discussed these issues in Chapter 4 and refers the reader to the National Research Council (1985b) report *Models for Biomedical Research: A New Perspective* and to the Office of Technology Assessment (1986) report *Alternatives to Animal Use in Research, Testing, and Education* for more information. Although recognizing that in many instances no alternatives may exist:

- The committee recommends that research investigators consider possible alternative methods before using animals in experimental procedures.

The National Library of Medicine, the National Agricultural Library, and other agencies have developed a number of databases and bibliographic resources for searching the broader sources of biological information.

- The committee recommends that databases and knowledge bases be further developed and made available for those seeking appropriate experimental models for use in the design of research protocols.

Much of the controversy surrounding animal experimentation is related to the use of animals from pounds. Many states and local communities have restricted the use of such animals. The effect of these restrictions has been to increase the cost of research for scientists who have relied on that source of animals.

Persons who object to the use of pound animals for research regard these animals as special because they may have been household pets. Those who believe that pound animals may be used point to the fact that over 10 million animals already are killed at pounds each year, precluding their use in adding to scientific knowledge.

- The committee unanimously recommends that pound animals be made available for research in which the experimental animals are used in acute experiments (in which the animals remain anesthetized until killed). While a majority of the committee supports the appropriate use of pound animals in all experiments, a minority opposes the use of pound animals for chronic, survival experiments.

References

Ames, B. N., J. McCann, and E. Yamasaki. 1975. Methods for detecting carcinogens and mutagens with the *Salmonella*/mammalian microsome mutagenicity test. Mutat. Res. 31:347–364.

Anderson, D., R. E. Billingham, G. H. Lampkin, and P. B. Medawar. 1961. The use of skin grafting to distinguish between monozygotic and dizygotic twins in cattle. Heredity 5:379.

Arnsten, A. F. T., and P. S. Goldman-Rakic. 1985. Alpha-adrenergic mechanisms in pre-frontal cortex associated with cognitive decline in aged non-human primates. Science 230:1272–1276.

Barlow, H. B. 1982. David Hubel and Torsten Wiesel—Their contributions towards understanding the primary visual-cortex. Trends Neurosci. 5:145–152.

Benaceraff, B. 1981. Role of MHC gene products in an immune regulation. Science 212:1229–1238.

Bieberich, C., and G. Scangos. 1986. Transgenic mice in the study of immunology. Bioessays 4:245–248.

Brent, L., C. G. Brooks, P. B. Medawar, and E. Simpson. 1976. Transplantation tolerance. Br. Med. Bull. 32:101–106.

Charpin, C., F. Kopp, N. Porreau-Schneider, J. C. Lissitsky, M. N. Lavant, P. M. Martin, and M. Toga. 1985. Laminin distribution in human decidua and immature placenta. Am. J. Obstet. Gynecol. 151:822–826.

Coalson, J. J., T. J. Kuehl, M. B. Escobedo, J. L. Hilliard, F. Smith, K. Meredith, W. Walsh, D. Johnson, D. Null, and J. L. Robotham. 1982. A baboon model of bronchopulmonary dysplasia: Pathologic features. J. Exp. Mol. Pathol. 37:335–350.

Cockburn, W. C., and S. G. Drozdov. 1970. Poliomyelitis in the world. Bull. WHO 42:405–406.

Cohen, B. J., and F. M. Loew. 1984. Laboratory animal medicine: Historical perspectives. Pp. 1–17 *in* Laboratory Animal Medicine, J. G. Fox, B. J. Cohen, and F. M. Loew, eds. Orlando, Fla.: Academic Press.

Cohen, C. 1986. The case for the use of animals in biomedical research. N. Engl. J. Med. 315:865–870.

Colloquium on Recognition and Alleviation of Animal Pain and Distress. 1987. J. Am. Veter. Med. Assoc. 191:1184–1298.

Coyle, J. T., D. L. Price, and M. R. de Long. 1985. Alzheimer's disease: A disorder of cortical cholinergic innervation. Pp. 418–431 *in* Neuroscience, P. H. Abelson, E. Butz, and S. H. Snyder, eds. Washington, D.C.: American Association for the Advancement of Science.

Deaton, J. G. 1974. New Parts for Old. Palisades, N.J.: Franklin.

DeBakey, M. E., and W. S. Henly. 1961. Surgical treatment of angina pectoris. Circulation 23(1):111–120.

De Kruif, P. H. 1926. Microbe Hunters. New York: Harcourt, Brace.

deLemos, R. A., D. M. Null, D. R. Gerstmann, T. J. Kuehl, and J. J. Coalson. 1985. Reduction of lung injury in prematures by use of HFOV. P. 213 *in* Proceedings of the 38th Conference on Engineering in Medicine and Biology, Chicago, September 30–October 2, 1985.

Deneau, G., Y. Tomoji, and M. H. Seevers. 1969. Self-administration of psychoactive substances by the monkey. Psychopharmacologia 16:30–48.

Doyle Dane Bernbach International Inc. 1983. Survey, Research and Marketing Services. New York: Doyle Dane Bernbach International Inc.

Draize, J. H., G. Woodward, and H. O. Clavery. 1944. Methods for the study of irritation and toxicity of substances applied topically to the skin and mucous membranes. J. Pharmacol. Exp. Ther. 82:377–390.

Eccles, J. C. 1957. The Physiology of Nerve Cells. Baltimore, Md.: Johns Hopkins University Press.

Eccles, J. C., and J. J. Gibson. 1979. Sherrington: His Life and Thought. Berlin: Springer-Verlag.

Enders, J. F., T. H. Weller, and F. C. Robbins. 1949. Cultivation of the Lansing strain of poliomyelitis virus in cultures of various human embryonic tissues. Science 109:85–87.

Ephrussi, B. 1942. Chemistry of eye color hormones of *Drosophila*. Q. Rev. Biol. 17:327–338.

Escobedo, M. B., J. L. Hilliard, F. Smith, K. Meredith, W. Walsh, D. Johnson, D. Null, J. J. Coalson, T. J. Kuehl, and J. L. Robotham. 1982. A baboon model of bronchopulmonary dysplasia: Clinical features. J. Exp. Mol. Pathol. 27:323–334.

Etter, J. C., and A. Wildhaber. 1985. Biopharmaceutical test of ocular irritation in the mouse. Food Chem. Toxicol. 23:321–323.

Flexner, S., and P. A. Lewis. 1909. The transmission of acute poliomyelitis to monkeys. J. Am. Med. Assoc. 53:1639.

Foundation for Biomedical Research. 1987. The Use of Pound Animals in Biomedical Research and Education. Washington, D.C.: Foundation for Biomedical Research.

Freiherr, G. 1986. Recombinant DNA research and wildlife studies focus on reducing the spread of rabies. Res. Resources Report (NIH) 10(4):1.

French, R. D. 1975. Antivivisection and Medical Science in Victorian Society. Princeton, N.J.: Princeton University Press.

Gardner, R. A., and B. T. Gardner. 1969. Teaching sign language to chimpanzees. Science 165:664–672.

Gay, W. I. 1984. The dog as a research subject. Physiologist 27:133–140.

Gay, W. I. 1986. Health Benefits of Animal Research. Washington, D.C.: Foundation for Biomedical Research.

Gibbs, C. J., A. Joy, R. Heffner, M. Franko, M. Miyazaki, D. M. Asher, J. E. Parisi, P. W. Brown, and D. C. Gadjusek. 1985. Clinical and pathological features and laboratory confirmation of Creutzfeld-Jakob disease in a recipient of pituitary-derived Human Growth Hormone. N. Engl. J. Med. 313:731.

Gloxhuber, C. 1985. Modification of the Draize eye test for the safety testing of cosmetics. Food Chem. Toxicol. 23:187–188.

Gorman, C. 1988. When guinea pigs become patients. Time, January 11, p. 67.

Hamburg, D. A., G. R. Elliott, and D. L. Parron, eds. 1982. Health and Behavior: Frontiers of Research in the Biobehavioral Sciences. Washington, D.C.: National Academy Press.

Hampson, J. 1985. Laboratory Animal Protection Laws in Europe and North America. Causeway, England: Royal Society for the Prevention of Cruelty to Animals.

Hand, L. 1960. The Spirit of Liberty; Papers and Addresses of Learned Hand. Chicago: University of Chicago Press.

Hodgkin, A. L., and A. F. Huxley. 1952. A quantitative description of membrane current and its application to conduction and excitation in nerve. J. Physiol. 117:500–544.

Hubbard, W. N., Jr. 1976. The origins of medicine. Pp. 685–721 *in* Advances in American Medicine: Essays at the Bicentennial, J. Z. Bowers and E. F. Purcell, eds. New York: Josiah Macy, Jr., Foundation.

Hume, D. M., J. P. Miller, B. F. Miller, and G. W. Thorn. 1955. Experiences with renal homotransplantation in humans: Report of nine cases. J. Clin. Invest. 34:327.

Institute of Medicine. 1986. Confronting AIDS: Directions for Public Health, Health Care, and Research. Washington, D.C.: National Academy Press.

Johnsen, D. O., and L. A. Whitehair. 1986. Research facility breeding. Pp. 499–511 *in* Primates: The Road to Self-Sustaining Populations, K. Benirschke, ed. New York: Springer-Verlag.

Kandel, E. R., and J. H. Schwartz, eds. 1985. Principles of Neural Science, 2nd ed. New York: Elsevier.

Kupiec-Weglinski, J. W., M. A. Filho, T. B. Strom, and N. L. Tilney. 1984. Sparing of suppressor cells: A critical action of cyclosporin. Transplantation 38:87–101.

Landsteiner, K., and H. Popper. 1909. Z. Immunitätsforsch. 2:377.

Lave, L. B., and G. S. Omenn. 1986. Cost-effectiveness of short-term tests for carcinogenicity. Nature 324:29–34.

Leader, R. W., and D. Stark. 1987. The importance of animals in biomedical research. Perspect. Biol. Med. 30:470–485.

Lillie, F. R. 1917. The freemartin; a study of the action of sex hormones in the foetal life of cattle. J. Exp. Zool. 23:371.

Loew, F. M. 1982. Developments in the history of the use of animals in medical research. Pp. 3–6 *in* Scientific Perspectives on Animal Welfare, W. J. Dodds and F. B. Orlans, eds. New York: Academic Press.

Luepke, N. P. 1985. Hen's egg chorioallantoic membrane test for irritation potential. Food Chem. Toxicol. 23:287–291.

McGrew, R. E. 1985. Encyclopedia of Medical History. New York: McGraw-Hill.

Medawar, P. G. 1944. Behaviour and fate of skin autografts and skin homografts in rabbits. J. Anat. 78:176.

Media General. 1985. Are Laboratory Animals Treated Humanely? New York: Associated Press.

Miller, N. E. 1985. The value of behavioral research on animals. Am. Psychol. 40:423–440.

Morowitz, H. J. 1986. Myasthenia gravis and arrows of fortune. Hosp. Pract. 21:179–194.

Moule, Y. 1984. An approach to the detection of environmental tumor promoters by a short-term cultured-cell assay. Food Addit. Contam. 1:199–203.

National Association for Biomedical Research. 1987. State Laws Concerning the Use of Animals in Research. Washington, D.C.: National Association for Biomedical Research.

National Institute of Drug Abuse. 1984. Research Monograph 52. *In* Testing Drugs for Physical Dependence and Potential and Abuse Liability, J. V. Brady and S. E. Lukas, eds. Bethesda, Md.: The Committee on Problems of Drug Dependence, Inc.

National Institutes of Health. 1981. Trends in Bioassay Methodology: In Vivo, In Vitro and Mathematical Approaches. National Institutes of Health Publ. No. 82-2382. Washington, D.C.: U.S. Department of Health and Human Services.

National Institutes of Health. 1987. NIH Centennial Media Kit: NIH's Role in Protecting Animals in Research. Bethesda, Md.: National Institutes of Health.

National Research Council. 1977. The Future of Animals, Cells, Models, and Systems in Research, Development, Education, and Testing. Proceedings of a symposium sponsored by the Institute of Laboratory Animal Resources. Washington, D.C.: National Academy Press.

National Research Council. 1980. National Survey of Laboratory Animal Facilities and Resources. A report of the Institute of Laboratory Animal Resources Committee on Laboratory Animal Facilities and Resources. National Institutes of Health Publ. No. 80-2091. Washington, D.C.: U.S. Department of Health and Human Services.

National Research Council. 1985a. A Guide for the Care and Use of Laboratory Animals. A report of the Institute of Laboratory Animal Resources Committee on Care and Use of Laboratory Animals. National Institutes of Health Publ. No. 85-23. Washington, D.C.: U.S. Department of Health and Human Services.

National Research Council. 1985b. Models for Biomedical Research: A New Perspective. A report of the Board on Basic Biology Committee on Models for Biomedical Research. Washington, D.C.: National Academy Press.

New York Academy of Sciences. 1988. Interdisciplinary principles and guidelines for the use of animals in research, testing, and education. A report of the Ad Hoc Committee on Animal Research. New York: New York Academy of Sciences. 26 pp.

Norton, S. 1981. Behavioral evaluation of nervous system damage in mammals and birds. Pp. 99–112 *in* Trends in Bioassay Methodology. National Institutes of Health Publ. No. 82-2382. Washington, D.C.: U.S. Department of Health and Human Services.

Office of Technology Asessment. 1986. Alternatives to Animal Use in Research, Testing, and Education. OTA Publ. No. OTA-BA-273. Washington, D.C.: U.S. Government Printing Office.

Orlans, F. B. 1987. Effective institutional animal care and use committees. Anim. Sci. 37(7, Supplement).

Paton, W. 1984. Man & Mouse: Animals in Medical Research. Oxford, England: Oxford University Press.

Pavlov, I. P. 1927. Conditioned Reflexes. (G. V. Anrep, trans.) London: Oxford University Press.

Pedersen, N. C., E. W. Ho, M. L. Brown, and J. K. Yamamoto. 1987. Isolation of a T-lymphotropic virus from domestic cats with an immunodeficiency-like syndrome. Science 235:790–793.

Public Health Service. 1985. Public Health Service Policy on Care and Use of Laboratory Animals. ILAR News 28:8–13.

Regan, T. 1983. The Case for Animal Rights. Berkeley: University of California Press.

Research Strategies Corp. 1985. Members of the American Public Comment on the Use of Animals in Medical Research and Testing. A Study Conducted for the Foundation for Biomedical Research. Princeton, N.J.: Research Strategies Corp. 119 pp.

Romski, M. A., R. A. White, C. E. Miller, and D. M. Rumbaugh. 1984. Effects of computer-keyboard teaching on the symbolic communication of severely retarded persons: Five case studies. Psychol. Rec. 34:39–54.

Rowan, A. N. 1984. Of Mice, Models, and Men: A Critical Evaluation of Animal Research. Albany: State University of New York Press.

Rowan, A. N., and A. M. Goldberg. 1985. Perspectives on alternatives to current animal testing techniques in preclinical toxicology. Annu. Rev. Pharmacol. Toxicol. 25:225–247.

Russell, W. M. S., and R. L. Burch. 1959. The Principles of Humane Experimentation Techniques. London: Methuen.

Salk, J. 1983. The virus of poliomyelitis. From discovery to extinction. J. Am. Med. Assoc. 250:808–810.

Sallach, H. J. 1972. Intermediary Metabolism Charts. Milwaukee, Wis.: P. L. Biochemicals, Inc.

Seevers, M. H. 1968. Psychopharmacological elements of drug dependence. J. Am. Med. Assoc. 206:1263–1266.

Singer, P. 1975. Animal Liberation. New York: Avon Books.

Skinner, B. F. 1938. The Behavior of Organisms. New York: Appleton-Century.

Society of Toxicology. 1986. Position Statement Regarding the Use of Animals in Toxicology. Washington, D.C.: Society of Toxicology.

Sokol, H. W., and H. Valtin, eds. 1982. The Brattleboro rat. Ann. N.Y. Acad. Sci. 394:1–828.

Starzl, T. E., and J. H. Holmes. 1964. Experience in Renal Transplantation. Philadelphia: W. B. Saunders.

Stone, L. 1977. The Family, Sex and Marriage in England 1500–1800. London: Weidenfeld and Nicholson.

Tennant, R. W., B. H. Margolin, M. D. Shelby, E. Zeiger, J. K. Haseman, J. Spanding, W. Caspary, M. Resnick, S. Stasiewicz, B. Anderson, and R. Minor. 1987. Prediction of chemical carcinogenicity in rodents from in vitro genetic toxicity assays. Science 236:933–941.

Theta Corporation. 1986. A Research Animal Market Report. No. 597. Middlefield, Conn.: Theta Corporation.

Thorndike, E. L. 1898. Animal intelligence: An experimental study of the associative processes in animals. Psychol. Rev. Monogr. 8 (Suppl. 2).

Turner, J. 1980. Reckoning with the Beast: Animals, Pain, and Humanity in the Victorian Mind. Baltimore, Md.: Johns Hopkins University Press.

U.S. Department of Agriculture, Animal and Plant Health Inspection Service. 1972–1987. Animal Welfare Enforcement: Report of the Secretary of Agriculture to the President of the Senate and the Speaker of the House of Representatives. Washington, D.C.: U.S. Government Printing Office.

U.S. Department of Agriculture. 1985. Agricultural Statistics. GPO No. 001-000-044010. Washington, D.C.: U.S. Department of Agriculture.

U.S. National Center for Health Statistics. 1988. Vital Statistics of the United States, 1985. Life Tables, Vol. II, Sect. 6. PHS No. 881104. Washington, D.C.: U.S. Government Printing Office.

Wiesel, T. N. 1982. The postnatal development of the visual cortex and the influence of environment. Nature 299:583–592.

Willis, W. D. 1985. The pain system: The neural basis of nociceptive transmission in the mammalian nervous system. *In* Pain and Headache, Vol. 8, P. L. Gildenberg, ed. Basel: Karger.

Yoshioka, Y., Y. Ose, and T. Sato. 1985. Testing for the toxicity of chemicals with *Tetrahymena pyriformis*. Sci. Total Environ. 43:149–157.

Individual Statements by Members of the Committee on the Use of Laboratory Animals in Biomedical and Behavioral Research

These individual statements appear exactly as the committee members prepared them. The National Research Council neither endorses nor takes responsibility for the content of the statements.

ARTHUR C. GUYTON

This statement is made for two purposes: first, to express severe disappointment that our Committee Report fails to make clear how seriously the Animal Rights Movement and increasing government regulation are impeding essential medical research; and, second, to record at least one dissenting vote against the implication in the "Recommendations" section of the main report that the present regulatory framework will allow a healthy future for medical research.

The success of the Animal Rights Movement in making medical research difficult has been phenomenal in the last 3 years. One-fifth of all States have already passed laws prohibiting release of pound animals for medical research. And multiple animal rights-welfare organizations have announced publicly their priority goal to eliminate by law all release of pound animals for medical research within the next few years. Historically, most large-animal medical research has been performed in dogs and cats obtained from pounds because these are all *unwanted* animals and because the cost to society in using these animals is almost zero, which contrasts with a cost of many millions of dollars when alternative animals are used.

Also, the Animal Rights Movement has been surprisingly effective in getting the Federal Government to establish very restrictive regulations on medical research. Some of the most blatant of these are: 1) The necessity to obtain prior approval before performing each type of animal experiment, a requirement that often delays essential research as much as two months. 2) A requirement that all major survival surgery on rabbits or larger animals be performed in a surgical operating room suite costing an average of a quarter million dollars and directed by a high-salaried veterinarian, even though the veterinarian usually is not a trained surgeon. In the past, this type of surgery has been done exceedingly successfully in the investigator's own laboratory at no extra cost. 3) Very arbitrary regulations for specific cage sizes, and even these have been changed on multiple occasions, costing hundreds of millions of dollars throughout the United States. These are only examples of a litany of such regulations.

The net effect has been an extreme increase in the cost of animals used in research as well as cost of lost time and effort by the investigator. Including the expense of meeting federal regulations, the cost of dogs and cats used in research, together with the cost of their care, now averages more than $1,000 per animal in some institutions, and *this does not count the cost of the research itself.* Historically, when animals were readily available on a day's notice from local animal pounds, the cost of dogs and cats was very little.

Role of Veterinarian Professionalism in Imposing New Difficulties for Medical Research. Veterinarian scientists have made and are making major contributions to medical research. However, in the last three years, there has been a proliferation of new government regulations requiring vastly expanded and costly roles for veterinarians as regulators of virtually all animal-based biomedical research. This presumably has come about because those government agencies that make the regulations (for example, the Inspection Agency of the Department of Agriculture) are staffed to a great extent by professional veterinarians, and they naturally believe that others cannot have the expertise to work properly with animals. Yet, we all know that medical professionalism, with doctors regulating doctors, and legal professionalism, with lawyers regulating lawyers, always under the pretense of high principles, make medical and legal services extremely expensive to the public. In a similar manner, this new proliferation of animal-control regulations requires a very costly layer of veterinarian regulators who do not actually participate in the research itself. The vast and superb medical research accomplishment of the past has been achieved without this new bureaucracy. Is it truly needed now? And if so, is it not also needed for the pounds and animal rights-welfare shelters which house and kill 50 times as many dogs and cats each year as does medical research?

Misplaced Faith in "Alternatives" to Animal Research. The Committee Report contains an entire chapter on *Alternative Methods in Biomedical and Behavioral Research.* Unfortunately, the prominence of this chapter gives false hope that animal-based medical research can be done with substitutes for animals. Unless we substitute human beings as the research subjects, this is very rarely true. Therefore, it is seriously wrong for the Committee Report to give such false expectation. The Animal Rights Movement has already made a strong effort in Congress to divert as much as one-fifth to one-half of all health-related research money to studies using only animal "alternatives," and our report will likely be used to support further such efforts.

Desperate Need for Help in Combating the Initiatives of the Animal Rights Movement and of Regulatory Bureaucracies. It is clear that the Animal Rights Movement, with the help of new and expanding federal, state, and local laws, is rapidly making much animal research cost ineffective as well as extremely wasteful of the research scientist's time. Many of the regulations appear not to have been thought through, such as the requirement for a quarter of a million dollar operating room suite to perform operations on rabbits.

The new federal regulations are similar to those established in Europe several decades ago; large animal research is now close to annihilation in Europe. As a result, the Europeans have made very little contribution in certain types of medical research, for example in cardiovascular surgery, except when the research could be done on human beings themselves.

Therefore, the medical research community desperately needs strong help in combating both the Animal Rights Movement and the growing regulatory bureaucracies. Our committee has failed to produce a document that will be helpful for this purpose. This is understandable because the committee itself includes many members who have never worked in animal research and particularly includes two Presidents of national animal "welfare" organizations.

Considering the rapidly expanding restrictive and time-wasting regulatory environment, I cannot in all good conscience recommend to young researchers that they pursue careers in those types of medical research that require the use of animals. How will it be possible to make many new advances in medicine?

CHRISTINE STEVENS

The report refuses to face the widespread, ingrained problem of unnecessary suffering among the millions of laboratory animals used yearly in our country, nor does it make so much as a passing reference to the serious problem of poor research using excessive numbers of animals.

The single recommendation, approved by majority vote, to improve the treatment of about 85% of research animals was reversed at the only Committee meeting I did not attend. The reversed recommendation requested the Secretary of Agriculture to issue regulations under the Animal Welfare Act extending its protection to mice, rats, birds and farm animals used for biomedical research.

Ironically, for lack of application of the minimum standards of the Animal Welfare Act, conditions of extreme neglect and abuse developed in a rodent laboratory under the jurisdiction of a Committee member. Dozens of photographs documenting long-standing filth, holes in walls and roof through which wild rodents gain access, hazardous handling of carcinogens and other improprieties which could confound test results and endanger personnel, were sent me by a concerned worker who asked my help in obtaining desperately needed reforms.

Another member of the Committee indicated that his institution doesn't know the number of mice and rats used and if reported, it is not the truth.

The report was to have provided new factual information on numbers of animals used, but the study was never conducted. Thus, there is no new quantification on animal use as announced by NAS when the Committee was formed. Readers are led to believe that animal use, especially of primates, is declining, e.g. "The substantial decrease (47 percent) in the use of nonhuman primates..." (p. 61). But USDA figures document an increase of 26.48% from 1986 to 1987. Total animal use also increased as did animals reported by the institutions as experiencing unrelieved pain. The chart (pp. 20–21) omits available USDA data on wild animals.

Although the report claims to favor animal welfare and oppose animal rights, the net effect is a *de facto* undermining of animal welfare.

I was shocked by the attitude of Committee members who asserted that we have no moral obligation to animals and expressed hatred of the idea of having a report that puts emphasis on alternatives. Committee members decried the public notion that animals have rights. If they do, one observed, I don't think we have the right to do animal experiments. During a discussion of current NIH regulations requiring that grant proposals provide data that will advance knowledge of immediate or potential benefit to humans and animals, members asked one another whether they agreed. We agree or we don't get any money was the response. It was surprising to hear the assertion that everybody cheats and prevaricates.

Although it is well known and widely acknowledged that the health and welfare of experimental animals is essential if sound observations are to be achieved, Committee members insisted that animal welfare rules should not be seen as a method of improving science.

Material presented by Committee members on the benefits of regulation of animal experimentation and the history of such regulation in Europe was cut from the report which instead makes the unreferenced charge that in "some countries" unspecified "strict legislation" has "reduced potential contributions to human welfare."

The modest U.S. legislation is unreasonably characterized as "a large body of laws and regulations" by which institutions and investigators are said to be "unnecessarily confused and burdened." Regulations under the 1985 amendments to the Animal Welfare Act have been held up by the very people who want to claim that "humane care and use of laboratory animals characterize the scientific community."

Virtually no acknowledgement of outstanding research results from scientific work appears in the report *unless* they were based on the use of vertebrate animals. Yet:

— a substantial proportion of NIH funds are dispensed for epidemiological and clinical research

— much animal experimentation produces no significant results

— leading scientists have publicly criticized erroneous conclusions resulting from large-scale animal experiments.

These exemplify matters on which readers of the report should receive objective information. But objectivity is incompatible with the "strong, hard-hitting report" promoters of animal experiments demand.

To prevent cruelty and theft by dog dealers and to encourage painless experiments in place of painful ones, I recommended "That dogs and cats obtained from public pounds be 1) used only for non-survival experiments under full anesthesia in which the animal is first rendered unconscious and never allowed to recover consciousness but passes directly into death, and 2) obtained directly by the registered research facility from the pound, not through a dealer." The recommendations on this subject (pp. 10, 73) do not accurately represent my proposal or my position. The report fails to take account of animal fear and pain.

The assertion that "most research animals are humanely killed at some point" is unreferenced. Not surprisingly, since there is no reporting system in place which would enable this assertion to be documented. We do not know how many animals are 1) killed humanely, 2) killed inhumanely, or 3) left to die unattended without pain relief.

The executive summary (p. 6) erroneously states that according to law: "... all animals used receive adequate presurgical and post-surgical care and pain-relieving drugs." But 130,373 animals were denied pain-relieving drugs under the law's exemption provision during 1987, according to the annual reports submitted by Registered Research Facilities to USDA.

Nothing in the report even hints at the long-drawn-out pain and suffering undergone by many laboratory animals. Instead, they are characterized as "minor" (p. 63) and a false claim is made that all serious violations have resulted in suspension of funding and/or imposition of fines. Mundane facts revealed in inspection reports of major research facilities by veterinary inspectors of the USDA are ignored. Typical findings of inspectors include:

- most rabbits without water
- excessive build up of manure and hair
- overcrowding
- moldy feed
- dogs with distemper
- piles of rodent droppings throughout building
- dogs standing in water
- rat holes numerous
- phenomenal number of roaches

— surgical site for implantation of electrodes into primates' brains conducted in office off busy hallway, only chemical sterilization of equipment

— monkeys wet and smeared with excreta

— dog needs resuturing

— sick kitten not under care of veterinarian. Blood from rectum and paresis of rear limb

— dogs sitting in urine and feces.

Such conditions for animals are a source of uncontrolled variables that skew research results, thus wasting scientific effort and taxpayers' money. The only acknowledgment of this suffering is one sentence: "From time to time some few members of the scientific community have been found to mistreat or inadequately care for research animals." The Executive Summary even waters down this feeble statement by omitting the word "mistreat."

A balanced report should recognize the severity and extent of the problem. It should recognize the essential role of sound regulation to prevent neglect and abuse of research animals, for the animals' sake and research accuracy. It should vigorously advocate:

1) research and development of alternatives, i.e., methods to reduce, refine or replace animal tests*

2) training laboratory personnel in humane care and treatment of animals

3) choice of least painful procedures by investigators

4) substantial government funding for data bases designed to: reduce unintended duplication of animal tests, facilitate distribution of information on alternatives and make non-warm-blooded animal systems available to investigators and students.

---

*An earlier version stated "Mammalian usage can be decreased by various techniques of replacement and reduction... More such models will become available particularly if additional research is devoted to this effort."

Appendix A
1896 Report of the National Academy of Sciences

In 1896, the National Academy of Sciences was asked to "express an opinion as to the scientific value of experiments upon the lower animals and as to the probable effect of restrictive legislation upon the advancement of biological science." The request was initiated by Senator Jacob Gallinger in response to a request from the surgeon generals and the Chief of the Bureau of Animal Industry. Gallinger had introduced legislation to restrict the use of animals in research. The request was as follows:

To the Hon. Jacob H. Gallinger, Senator of the United States,
Chairman of Subcommittee, etc.

Washington, D.C., April 24, 1896.

Sir: Referring to Senate bill 1552, we respectfully invite your attention to the fact that the National Academy of Sciences is now in session in this city, and that this body is generally recognized as the highest scientific tribunal in the United States; also that the act incorporating it contains the following clause: "And the Academy shall, whenever called upon by any Department of the Government, investigate, examine, experiment, and report upon any subject of science or art." (Act approved March 3, 1863).

We respectfully request that the National Academy of Sciences be called upon to express an opinion as to the scientific value of experiments upon the lower animals and as to the probable effect of restrictive legislation upon the advancement of biological science.

Very respectfully,

D.E. Salmon,
Chief of the Bureau of Animal Industry.
J.R. Tryon,
Surgeon General U.S.N.
Geo. M. Sternberg,
Surgeon General U.S.A.
Walter Wyman,
Surgeon General United States Marine Hospital Service.

The National Academy of Sciences' response was drafted by Harvard physiologist H.P. Bowditch and conveyed to Senator Gallinger in a letter from Academy President Wolcott Gibbs. The response was as follows:

To the Hon. Jacob H. Gallinger.

Washington, D.C., April 24, 1896

Sir: I have the honor to acknowledge the receipt of a letter addressed to you by D.E. Salmon, the Chief of the Bureau of Animal Industry; J.R. Tryon, Surgeon General United States Navy; George M. Sternberg, Surgeon General United States Army; and Walter Wyman, Surgeon United States Marine Hospital Service, asking that the National Academy of Sciences be called upon to express an opinion as to the scientific value of experiments upon the lower animals and as to the probable effect of restrictive legislation upon the advancement of biological science. The letter of these gentlemen is supplemented by an expression of your desire that the National Academy of Sciences should report or make suggestions upon the subject. In accordance with your request, I have the honor to submit to you the following report as the expression of the opinion of the National Academy of Sciences:

Biology is the science of living organisms and tissues, and must therefore advance by means of observations and experiments made upon living beings. One of its most important branches, viz, physiology, or the science which deals with all the phenomena of life, from the activity of bacteria to that of the brain cells of man, forms the foundation upon which the science and practice of medicine are built up, since a knowledge of the bodily functions in their normal state is essential for the understanding and treatment of those derangements of function which constitute disease.

The fact that the pursuit of physiology consists chiefly in the study of physical and chemical phenomena as manifested by living beings makes it necessary that physiology should be studied by experimental methods. The physiologist, no less than the physicist and the chemist, can expect advancement of his science only as the result of carefully planned laboratory work. If this work is interfered with, medical science will continue to advance, as heretofore, by means of experiment, for no legislation can affect the position of physiology as an experimental science; but there will be this important difference, that the experimenters will be medical practitioners and the victims human beings.

That animals must suffer and die for the benefit of mankind is a law of nature, from which we can not escape if we would, and as long as man claims dominion over the brute creation and asserts his right to kill and mutilate animals in order to obtain food and clothing, and even for purposes of amusement and adornment, it is surely unreasonable to wage a humanitarian warfare against the only kind of pain-giving practice that has for its object the relief of pain.

The death of an animal in a physiological laboratory is usually attended with less suffering than is associated with so called natural deaths, for the discovery of anaesthetics has extended its beneficent influence over the lower animals as well as over the human race, and in modern laboratories anaesthetics are always employed, except when the operation involves less suffering to the animal than the administration of the anaesthetic (as in the case of inoculations) or in those rare instances in which the anaesthetic would interfere with the object of the experiment. The suffering incident in biological investigation is therefore trifling in amount and far less than that which is associated with most other uses which man makes of the lower animals for purposes of business or pleasure.

As an offset to this trifling amount of animal suffering are to be placed incalculable benefits to the human race. From the time when Aristotle first discovered the insensibility of the brain to the time when the latest experiments in the use of antitoxin have largely robbed diphtheria of its terror, almost every important advance in the science of medicine has been the direct or the indirect result of knowledge acquired through animal experimentation.

It is, of course, conceivable that persons whose occupations lead them to sacrifice animal life for scientific purposes may at times pay too little regard to the suffering which they inflict, but the Academy understands that even those who advocate restrictive legislation by Congress do not claim that such abuses exist in the District of Columbia, and until evidence of this sort is presented it would seem to be the part of wisdom to leave the regulation of research in the hands of the governing bodies of the institutions in which the work is going on. The men engaged in this work are actuated by motives no less humane than those which guide the persons who desire to restrict their action. Of the value of any given experiment and of the amount of suffering which it involves they are, owing to their special training, much better able to judge. When the men to whom the Government has intrusted the care of its higher institutions of research shall show themselves incapable of administering them in the interest of science and humanity, then, and not till then, will it be necessary to invoke the authority of the National Legislature.

I have the honor to be, sir, very respectfully, your obedient servant,

WOLCOTT GIBBS,
President of the
National Academy of Sciences.

Appendix B

Curricula Vitae of Committee Members

NORMAN HACKERMAN (*Chairman*), received a Ph.D. degree in chemistry from the Johns Hopkins University and has served as professor of chemistry and as president of both the University of Texas and Rice University. A member of the National Academy of Sciences, he has served on many boards and advisory committees, including the National Science Board. His research interests include the study of corrosion of metals and the surface chemistry of metals and oxides.

KURT BENIRSCHKE received an M.D. degree from the University of Hamburg. His past academic posts have been at the medical schools of Harvard and Dartmouth Universities. He currently holds appointments at the University of California at San Diego and at the San Diego Zoo. His research interests involve pathology and reproductive medicine.

MICHAEL E. DeBAKEY received an M.D. degree from Tulane University. In addition, he holds numerous honorary degrees from both U.S. and foreign universities. A member of the Institute of Medicine, he is the Chancellor of the Baylor College of Medicine in Houston. His special research interest is in cardiovascular surgery.

W. JEAN DODDS was awarded a D.V.M. degree at the Ontario Veterinary College in Guelph, Canada. She is now chief of the

Laboratory of Hematology of the New York State Department of Health at Albany. Dr. Dodds' research interests involve laboratory animal medicine, including studies of comparative hemostasis and thrombosis and comparative immunohematology.

EDWARD L. GINZTON received a Ph.D. degree from Stanford University, where he continued his career until joining Varian Associates where he served as president and chairman of the board. He is a member of both the National Academy of Sciences and the National Academy of Engineering. His research interest is in applied physics.

CARL W. GOTTSCHALK, who holds an M.D. degree from the University of Virginia, is a Career Investigator of the American Heart Association and is the Kenan Professor of Medicine and Physiology at the University of North Carolina. He is a member of both the Institute of Medicine and the National Academy of Sciences. His research specialty is renal physiology.

ARTHUR C. GUYTON was awarded an M.D. degree at Harvard Medical School. He currently serves as chairman of the Department of Physiology and Biophysics at the University of Mississippi School of Medicine. He has received many awards for his research in circulatory physiology and medical electronic development.

WILLIAM HUBBARD earned an M.D. degree from New York University. Following an academic career, which included serving as dean of the Medical School at the University of Michigan, he joined the Upjohn Company, where he still serves as president emeritus. He is a member of the Institute of Medicine, and his research interest lies in medical education.

JOHN KAPLAN received a law degree from Harvard University. He has held academic appointments at Northwestern University, the University of California at Berkeley, and Stanford University, where he is the Jackson Eli Reynolds Professor of Law. He has published books on a variety of today's concerns, including drug control and drug abuse.

HAROLD J. MOROWITZ was awarded a Ph.D. from Yale University, where he serves as professor of molecular biophysics and biochemistry. His research emphasizes the thermodynamic foundations of biology, the study of prebiotic chemistry, and the matrix of biological information.

CARL PFAFFMANN holds a Ph.D. degree from Cambridge University. Following various teaching positions at Brown, Yale, and Harvard Universities, he was appointed the Vincent and Brooke Astor Professor of Physiological Psychology, now Emeritus, at Rockefeller University. He is a member of the National Academy of Sciences, and his research interests are in neurophysiology and behavior.

DOMINICK P. PURPURA received an M.D. degree from Harvard University Medical School. He has held academic appointments at Columbia and Stanford Universities, as well as the Albert Einstein College of Medicine, where he now serves as dean. He is a member of both the National Academy of Sciences and the Institute of Medicine. His research interests are in the field of developmental neurobiology.

CHRISTINE STEVENS was educated at the University of Michigan and the Cranbrook Art Institute. She is the founder and president of the Animal Welfare Institute and has written numerous articles and edited books on the subjects of animal welfare and protective legislation.

LEWIS THOMAS earned his M.D. degree from Harvard University. He is currently president emeritus of Memorial Sloan-Kettering Cancer Center. He is the author of *Lives of a Cell*, as well as other popular books. He is a member of both the National Academy of Sciences and the Institute of Medicine. His research is centered on infectious diseases.

JAMES McKENDREE WALL was awarded a B.D. from Emory University and an M.A. in divinity from the University of Chicago. He has received several honorary degrees and is an ordained United Methodist minister. He has had a career in journalism and is currently editor of the *Christian Century*, published in Chicago.

Appendix C
Invited Speakers at Committee Meetings

Bonnie V. Beaver
College of Veterinary Medicine
Texas A&M University

Charles Cleveland
Science and Technology Division
Pharmaceutical Manufacturers Association
Washington, D.C.

J. Frederick Cornhill
Department of Surgery
Ohio State University Medical School

James Fox
Division of Comparative Medicine
Massachusetts Institute of Technology

Alan Goldberg
Center for Alternatives to Animal Testing
The Johns Hopkins School of Public Health

Barbara C. Hansen
Graduate Studies and Research
University of Maryland
Baltimore

Harold A. Hoffman
Animal Genetics Systems, Inc.
Rockville, Md.

J. Patrick Jordan
Cooperative State Research Service
U.S. Department of Agriculture
Washington, D.C.

David Kingsbury
Biological, Behavioral, and Social Sciences
National Science Foundation
Washington, D.C.

Judy Kosovich
Office of Technology Assessment
U.S. Congress
Washington, D.C.

Franklin M. Loew
School of Veterinary Medicine
Tufts University

Tom Regan
Department of Philosophy and Religion
North Carolina State University

Daniel H. Ringler
Department of Laboratory Animal Medicine
University of Michigan Medical School

Harry Rowsell
Canadian Council on Animal Care
Ottawa, Ontario, Canada

John H. Seamer
Humane Family Foundation
New Milford, Ct.

Richard Simmonds
Instructional and Research Support
Uniformed Services University of the Health Sciences

John L. VandeBerg
Department of Genetics
Southwest Foundation for Biomedical Research
San Antonio, Tex.

James Vorosmarti
U.S. Department of Defense
Washington, D.C.

James Willett
Division of Research Resources
National Institutes of Health
Bethesda, Md.

Index

B

C

E

F

G

S

T

U

V

Y